EXPLORING REALITY

Exploring Reality

The Intertwining of Science and Religion

JOHN POLKINGHORNE

Yale University Press   New Haven and London

Published with assistance from the Louis Stern Memorial Fund.

Scriptural quotations throughout this work are taken from the
New Revised Standard Version (NRSV).

Published 2005 in the United States by Yale University Press and in
Great Britain by SPCK.

Set in Janson type by Tseng Information Systems, Inc.
Printed in the United States of America.

Library of Congress Cataloging-in-Publication Data
Polkinghorne, J. C., 1930–
Exploring reality : the intertwining of science and religion /
John Polkinghorne.
p.   cm.
Includes bibliographical references and index.
ISBN-13: 978-0-300-11014-2 (alk. paper)
ISBN-10: 0-300-11014-6
1. Religion and science.   2. Apologetics.   3. Reality.   I. Title.
BL240.3.P64   2005
261.5′5—dc22   2005012580

A catalogue record for this book is available from the British Library.

The paper in this book meets the guidelines for permanence and durability
of the Committee on Production Guidelines for Book Longevity of the
Council on Library Resources.

10 9 8 7 6 5 4 3 2 1

To my colleagues in the International Society
for Science and Religion

Contents

Preface

After twenty-five years working as a theoretical physicist, I decided that the time had come to do something else, so I turned my collar round and became an Anglican priest. For the past twenty years I have been a scientist-theologian, seeking to combine the perspectives of science and Christianity into a stereoscopic world view.

I have always wanted to make it clear that I did not leave physics because of any disillusionment with that subject. I retain a lively interest in science and a deep respect for all that it can tell us. Yet its enthralling account is not sufficient by itself to quench our thirst for understanding, for science describes only one dimension of the many-layered reality within which we live, restricting itself to the impersonal and general, and bracketing out the personal and unique. Many things altered in my life when I changed from being a physicist and became a priest, but one significant thing remained the same: the central importance of the search for truth. All my life I have been trying to explore reality. That exploration includes science, but it also necessarily takes me beyond it.

The process of investigation has a spiral character, as tackling the issues draws the explorer inwards to a deeper engagement with the multidimensional character of reality. This book reports the most recent cycle of my own exploratory thinking, presenting the latest communiqué from a continuing survey of the frontier between science and religion. Some of its themes are ones that I have visited before, but this compact volume aims to offer additional insight and I have refrained from undue repetition of what I have written already elsewhere. In consequence, there are some issues concerning which I am content simply to offer references to previous writings for the sake of a reader who wishes to follow up these points. Yet I believe that what is presented here is sufficiently self-contained for the volume to be read satisfactorily on its own. I suppose that the kind of iterative approach, pursued here and elsewhere in my work, is a pattern I inherited from all those years as an elementary particle physicist. In that subject, one progressed step by step, slowly building on what had gone before and concentrating at any one time on the pressing problems at hand.

I hope that this progress report will be of some help to others engaged in the great human quest for unified understanding. Because reality is multilayered, its exploration calls for a corresponding multiplicity of levels of enquiry. Account must be taken of what science, in its impersonal way, can tell us about the structure and process of the world, including a recognition that even within its own domain science cannot yet tell a fully integrated story. To this must be added due recognition of the dimension of the personal, acknowledging both the remarkable character of human nature and also the significance of its evolutionary origin. Unique resources for gaining deeper insight are offered to us by the lives and teaching of remarkable individuals. I am particularly concerned with the significance of Jesus of Nazareth. Testimony to meeting with the sacred reality of God, fostered and preserved within a faith tradition, is a dimension of human experience that must be taken with the utmost seriousness. For me as a Christian, this implies exploration of a trinitarian understanding of the divine nature. The first five chapters of the book cover this ground in a sequence with an intensifying focus, moving from the generality of science to the uniqueness of the one true God.

The remaining five chapters are concerned with issues that cluster round that opening progression. Human experience of time is fundamental to our encounter with reality, yet perplexing in its character. Honest recognition of the diversity of human religious perspectives, and admission of the apparent ambiguity of a world that is both beautiful and bitter in its nature, must both be part of engaging with reality as we actually encounter it. The moral dimension of human life, including contemporary perplexities that arise from seeking to apply ethical principles in novel and unprecedented situations, presents challenges to society in its search for right decisions and just procedures. All these issues have to be addressed.

The opening chapter is a very brief explanation of how natural the task of exploring reality is for someone whose intellectual formation has been in the sciences, notwithstanding the raised eyebrows of some postmodernist colleagues about any notion of access to the way things actually are. I nail my colours to the mast and assert my belief in critical realism. On this occasion, however, I do not seek to lay out the detailed defence that this philosophical stance undoubtedly requires, since I have written pretty extensively on this topic in other books of mine.

Chapter 2 surveys current understanding of the causal structure of the world. It centres on two main themes. One is that scientifically our knowledge is still pretty patchy, excellent within certain well-defined domains but often unable to make satisfactory connections between different domains. The problematic of the relationship of quantum physics to classical physics provides an instructive example. The second theme is that matters of causality, though certainly influenced by scientific discoveries, are not finally settled by science alone. Ultimate conclusions have to rest on the foundation of a metaphysical decision. Quantum theory, and chaos theory, and the issue of the relationship between them, all provide illustrative examples of this fact. This is the chapter that will make the greatest demands on the reader without much of a scientific background, but I have tried not to use ideas without offering a non-technical explanation of them. I believe that it is worth putting in some effort to be able to appreciate how the scientific account of causal structure, though deeply insightful, is still far from being complete.

Chapter 3 acknowledges the permanent power of the contribution made by evolutionary thinking to our understanding of the nature of reality. However, the success of evolutionary explanation requires an adequate characterisation of the environmental context within which the processes take place, as well as an understanding of the processes themselves. This kind of thinking depends for its success as much upon getting the ecology right as on getting the genetic factors right. Recognition of the remarkable richness of human experience, and of the unique abilities possessed by the genus *Homo*, persuades me that a simple neo-Darwinian account, based only on differential gene selection within an environment described solely in physico-biological terms, is inadequate to the task of understanding humanity. Human nature should certainly be considered in an evolutionary context, but this requires acknowledgement of the metaphysically rich setting in which that development has taken place. Noetic dimensions of reality, such as those of the mathematical and the moral, are as significant to the human story as are the dimensions of materiality. Human beings are psychosomatic unities living in a world whose understanding requires that a proper balance be struck between the mental and the material, an equalhandedness of the kind to which a philosophy of dual-aspect monism aspires. Human beings are also creatures who live in the veiled presence of their Creator. No account of human nature will be adequate that does not take the dimension of the sacred seriously. This requires an unfashionable acknowledgement of human heteronomy before God.

Chapter 4 turns from the generality of what has gone before to consideration of the astonishing influence of a specific individual. Jesus of Nazareth lived two thousand years ago in a peripheral province of the Roman Empire. He died a painful and shameful death, deserted by his followers and leaving no personally written account of himself or his ideas. On the face of it, it seems a story of initial promise ending in dismal failure. Yet Jesus has been one of the most influential figures in world history. This chapter approaches his life from the point of view of a careful historical evaluation of what can be learned about him from the gospel material. My conclusion, however, is that an adequate explanation of why the story of Jesus continued beyond the apparently miserable end of his crucifixion forces the enquirer to be prepared to go beyond the limited possibilities of an historical naturalism, predicated on the belief that what usually happens is what always happens. The chapter indicates the evidential motivations that point to taking seriously the claim of the truth of Jesus' resurrection, but it acknowledges that the way this evidence is weighed must depend upon theological judgements that go beyond the verdicts of purely secular history. My principal purpose in this chapter is to demonstrate that the question of Jesus is an indispensable item on the agenda for the exploration of reality, but I do not repeat the full defence of an orthodox Christian assessment of his significance that I have given already in my Gifford Lectures.

Chapter 5 turns instead to a topic on which I have written only sparingly before, trinitarian theology. Emphasis is laid on the manner in which trinitarian thinking arose as a response to Christian experience, both that recorded in scripture and that continued in the worshipping life of the Church, rather than from any unbridled indulgence in rash speculation about the ineffable nature of God. In recent years there has been a considerable revival of interest in the theological fruitfulness of trinitarian thinking. Much of the motivation for this has come from an enhanced recognition of the fundamental significance of relationality, a development that can draw some support from aspects of modern science. This is the chapter that will make the most demands on a reader without much of a theological background but, as in the case of science in chapter 2, I endeavour to explain the main concepts to which I refer.

Time remains a continuing topic for metaphysical discussion and dispute. Chapter 6 defends an unfolding and developmental view of temporality against the atemporal claims of the proponents of the block universe. The stance adopted carries with it implications for the Creator's relationship to the time of creation and for the character of divine omniscience. I also suggest that it encourages a developmental view of scripture, seen as the record of an evolving divine revelation conditioned at all stages by the contingencies of history.

One of the most perplexing problems for someone seeking to explore divine reality arises from the conflicting testimonies of the world faith traditions. While I write from within the Christian tradition, I cannot do so unaware of the different insights and understandings of my brothers and sisters in other religious communities. Chapter 7 acknowledges that the ecumenical encounter of the world religions must take place on the basis of recognising both the clashing diversity and the spiritual authenticity that are present in what the traditions have to say. I do not have much new to say on this important topic beyond emphasising that, from a Christian perspective, it is belief in the veiled working of the Spirit that provides the theological basis for undertaking what will undoubtedly prove to be a long and difficult dialogue.

Chapter 8 considers the perplexing challenge to theism presented by the existence of evil and suffering. It suggests that the free-will defence offers some insight into the presence of moral evil, and an analogous free-process defence into the presence of natural evil. It is acknowledged, however, that the challenge of evil cannot be met solely at the level of philosophical argument. The deepest Christian response lies in recognising the passion of Christ as being divine participation in the travail of creation, so that the crucified God is truly a fellow sufferer who understands.

Ethical issues cannot simply be treated in broad generality, for many of the greatest challenges that we face come from very specific problems and perplexities. During the past fifteen years I have had the privilege of serving on several committees charged with offering advice to the United Kingdom Government about ethical issues that have arisen from scientific and technological advances, particularly in the field of genetics. In chapter 9 I seek to use this experience as a resource for exploring the complex nature of debate about the nature of ethical reality. The chapter looks at some of the current moral questions related to embryonic stem cells and other new genetic developments. Evaluation of feasibility by the relevant specialists is an essential input, but the possibilities have then to be assessed within a broader context of responsibility. The voices of the experts and of representatives of the ethically concerned public must both be given an adequate hearing. As a consequence, discussion has to be temperate and interdisciplinary if the technological power afforded by scientific advance is to be used in ethically justified ways that succeed in accepting the good but refusing the bad.

The book ends with chapter 10, a short speculative excursus in which I allow myself imaginative liberty to explore certain aspects of the eschatological hope of a destiny beyond death. The aim is modest: the exploration of some conceptual possibilities, rather than the certain establishment of the details of the life of heaven. I see the discussion as analogous to the use of 'thought experiments' in science.

Acknowledgements

I am grateful to the editorial staff at SPCK and at Yale University Press for their work in preparing the manuscript for press. I particularly thank Simon Kingston for many helpful comments on an early draft of the manuscript. I also thank my wife Ruth for her assistance with correcting the proofs.

The International Society for Science and Religion was founded to bring together leading participants in the field, drawn as widely as possible from all parts of the world and from all faith traditions. I had the privilege and honour to be its Founding President and I dedicate this book to the fellow members of our Society.

John Polkinghorne
Queens' College,
Cambridge.

CHAPTER ONE

Reality?

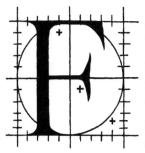 OR some the title of this book will be a red rag to a bull. They will dismiss it as exhibiting the author's naivety. 'Reality', and the closely-allied word 'truth', are not in common currency in some circles today, and consequently those who employ them lay themselves open to intellectual condescension and pity. I am unrepentant.

Much of the tone of contemporary sceptical discourse was already set by those nineteenth-century Masters of Suspicion, Friedrich Nietzsche and Sigmund Freud. The former once referred to truths as illusions that we have forgotten are illusions, and the latter, through his work in human psychology, suggested that the actual motivations for our beliefs often lie hidden in unconscious depths, so that they are frequently quite different from those which our conscious egos propose to us. Of course, each of these thinkers implicitly exempted their own ideas from subversion along the lines of their particular critiques, as also did Karl Marx in relation to the influence of economic factors and social class.

More recently, the extreme wing of the movement loosely categorised as postmodernism has suggested that instead of truth about reality, we have to settle for a portfolio of opinions expressing personal or societal points of view. Though there may appear to be conflicts between the different perspectives proposed, it is said that there is no real competition because, in fact, there is not actually anything to contend about. All points of view can claim equal authenticity, since none is constrained by an independently accessible external reality. The story goes that intellectual life is strictly à la carte.

Science has not been exempt from this assault on the possibility of rationally conclusive discourse. Its findings are held simply to be the products of the communities that propose them; its theorisings are supposed to be more about the exercise of power than about the attainment of veracity. For the extreme postmodernist, there are not really quarks and gluons as the constituents of matter, but the idea of them is a construct of the invisible college of physicists, who have simply colluded in seeing the world in a quarklike way.

As with many other reductive and dismissive accounts of human activity and human nature, these critiques are based at best on no more than quarter truths, whose scope is then exaggerated in the attempt to promote them into the pretension of total explanation. Of course, the motivations for human beliefs do lie at a variety of levels within the psyche, an insight known to Augustine and to generations of spiritual directors. Of course, scientific activity is influenced by cultural and social judgements of what investigations it would be valuable to pursue and viable to fund. Of course, experience has to be interpreted before it becomes truly interesting, and this introduces the danger of distortion through tricks of perspective, a problem that has to be recognised and taken into account. A naive objectivity of unproblematic 'facts' is far too crude a way to encapsulate our encounter with the way things are. Yet few critics of the ideas of truth and reality are so committed to that cause that it is matter of indifference to them what kind of doctor, witch or medical, they consult when they are ill. Nor do they tend to regard belief in the safe functioning of the aircraft they are about to board as being sufficiently established if it has arisen simply as the result of a socially negotiated consensus. At the beginning of the twentieth century, the positivist philosopher-physicist Ernst Mach denied the existence of atoms. Can any one really believe today that matter does not have an atomised structure? Scientific knowledge of a reliable kind really does increase. Of course, we know now that atoms themselves are made out of still smaller constituents (quarks, gluons and electrons). The maps that science makes of the physical world have always had to be open to revision when territory comes to be surveyed on a more intimate scale than had been explored hitherto. Yet these maps have proved reliable and trustworthy at the level of detail that they profess to describe. Science's achievement is not absolute truth, but it can rightly claim verisimilitude.

The realist counter-claim that is being made against the sceptics—a claim that certainly requires detailed defence—is that of a *critical realism*. The adjective is necessary because something more subtle than naive objectivity is involved (we do not see quarks directly, but their existence is indirectly inferred). The noun is justified because the best explanation of persistent scientific explanatory power and technological success is that science succeeds in describing, within the acknowledged limits of verisimilitude, the way things actually are.

Almost all scientists, consciously or unconsciously, are critical realists. Scientist-theologians are often self-confessed critical realists about both science and theology.[1] I have written rather often on the subject, seeking to base the argument on case studies, since I do not believe that it can be settled solely by abstract considerations.[2] I do not intend to repeat that discussion here. Let me be content to make three simple points:

(1) Defence of realism in science depends partly upon recognising the unexpected character often stubbornly displayed by nature. Far from its behaving like epistemological clay in our pattern-seeking hands, capable of being moulded into any pleasing shape that takes the fancy, the physical world frequently proves highly surprising, resisting our expectations and forcing us to extend, in unanticipated ways, the range of our intellectual understanding. In consequence, the feel of actually doing science is undeniably one of discovery, rather than pleasing construction. Theologians can claim something similar about the encounter with God. Time and again human pictures of deity prove to be idols that are shattered under the impact of divine reality.

1. For examples, see I. G. Barbour, *Myths, Models and Paradigms*, SCM Press, 1974; *Religion and Science*, SCM Press, 1998, ch. 5; A. R. Peacocke, *Intimations of Reality*, University of Notre Dame Press, 1984; *Theology for a Scientific Age*, SCM Press, 1993, pp. 7–23; and note 2.

2. J. C. Polkinghorne, *One World*, SPCK/Princeton University Press, 1986, chs 1–3; *Reason and Reality*, SPCK/Trinity Press International, 1991, chs 1 and 2; *Science and Christian Belief/The Faith of a Physicist*, SPCK/Fortress, 1994/1996, ch. 2; *Beyond Science*, Cambridge University Press, 1996, ch. 2; *Belief in God in an Age of Science*, Yale University Press, 1998, chs 2 and 5; *Faith, Science and Understanding*, SPCK/Yale University Press, 2000, chs 2, 3 and 5. 1.

(2) An experience fundamental to the pursuit of science is a sense of wonder, induced by the beautiful order and fruitful nature of the universe. There is an authenticity about science's discoveries of explanatory insight that is deeply persuasive that the scientists are 'onto something', gaining knowledge that comes from an external reality and which cannot be conceived as being simply an internally spun fable. Albert Einstein used often to express his awe at the order of nature, saying that he felt a mere child in the presence of the elders when confronted by such intellectual beauty. From deep simplicity comes immense complexity. For example, the genetic code, lying at the basis of all terrestrial life, depends upon certain chemical properties of the nucleotides forming DNA and RNA, which properties are themselves consequences of the outworking of the laws of electromagnetism and quantum theory. In a suitably compact notation, I could literally write the latter on the back of an envelope. No human ingenuity could be believed to be capable of constructing independently a story of such astonishing economy and fruitfulness. Its discovery required the genuine nudge of nature. The religious believer can find here grounds for understanding the universe as a creation, whose deep order and inherent fertility express the mind and will of its Creator.

(3) If interpreted experience is to be the basis of our understanding reality, then our concept of the nature of reality must be sufficiently extensive to be able to accommodate the richness of our experience. The many-levelled character of human encounter with the world resists all attempts to reduce it to a narrow account. In chapter 3 I shall discuss the context within which human life has evolved and is now lived and I shall argue that a just discussion requires recognition not only

of the physico-biological environment, but also of the realms of truth, beauty and goodness. There is an authenticity and richness in human life that demands that we take all of our experience with the utmost seriousness, respecting the multi-dimensional way in which it presents itself to us.

Any metaphysical world view that did not seek to take reality on reality's terms in the way that I have briefly sketched would be unacceptable. I want to use this short book to suggest some thoughts that arise from using trinitarian theology as a way of engaging with reality in the richness and variety of its actual impact upon us, resisting the temptation to embrace prematurely tidy schemes produced by those false reductions which, though they may sound speciously plausible in the abstraction of the study, deny the full authenticity of actual human experience, lived in the world.

CHAPTER TWO

The Causal Nexus of the World

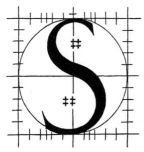 CIENCE achieves such frequent success in the many areas of its enquiry that it is difficult for us to remember how diverse are those areas and, in many ways, how little understood are the connections between them. Most scientists necessarily spend much of their time concentrating on their own specialised disciplines. Consequently, they seldom raise their eyes to look at the broader scene. Were they to do so, they would behold a fragmented picture, a patchwork of areas of insight only loosely, if at all, connected to each other.

One way of dividing up the scientific account is to introduce a hierarchy of forms of rational enquiry, ordered according to an ascending scale of the complexity of the entities under discussion. Proceeding in this way yields the canonical sequence: physics, chemistry, molecular biology, cellular biology, biology of organisms, neuroscience, psychology, an-

7

thropology, sociology.[1] Sub-sequences are readily identified within these levels of description. For example, in the case of physics there is a spectrum of internal complexity running from elementary particle physics to theories of condensed matter and continuum mechanics. The levels overlap to some degree (superconductors are surely more complicated than simple inorganic molecules, so physics and chemistry are not cleanly separated from each other) and the relationship between successive scientific levels can sometimes be a matter of uncertainty and dispute. While all would acknowledge that biochemistry throws very significant light on processes taking place within living entities, the complex character of even a single cell is such that it is by no means clear that a constituent account tells us all that we shall ever need to know about the richly complicated and integrated character of living entities. Francis Crick proclaimed that 'The ultimate aim of the modern movement in biology is in fact to explain *all* biology in terms of physics and chemistry',[2] but whether this ambitious programme is feasible or well conceived is open to very serious question.

Those who think like Crick are strong reductionists. For them, science's technique of splitting entities into constituent parts is not simply a convenient methodological strategy for tackling certain problems, but it also corresponds to the ontological character of nature itself. In their opinion, the constituent picture is simply the fundamental way things are, so

1. For more detailed hierarchical analyses, see A. R. Peacocke, *Theology for a Scientific Age*, SCM Press, 1993, pp. 212–48; N. Murphy and G. F. R. Ellis, *The Moral Nature of the Universe*, Fortress, 1996, ch. 2.

2. F. Crick, *Of Molecules and Men*, University of Washington Press, 1966, p. 10.

that the true account of reality lies solely at the lowest level, with the other levels in the hierarchy of complexity being just complex corollaries of what lies beneath. Logically this should lead these reductionists to accord the palm to elementary particle physics, and there are certainly some people in my old subject who are bombastic enough to entitle its still-unfulfilled quest for a Grand Unified Theory, the search for a 'Theory of Everything'. Yet strong reductionists often display a reluctance to sink below the level of their own discipline, so that geneticists want to attach special significance to genes,[3] and molecular biologists to molecules.[4]

Against the strong reductionists are the emergentists, whose slogan is 'More is different'. For them the whole exceeds the sum of its parts, so that it would be absurdly inappropriate to call a constituent account a Theory of Everything. They point to the degree of conceptual independence that exists between the various levels of the hierarchy of sciences. It is clear that the fitness of an organism for survival in an ecological setting is not an idea that can usefully be transcribed into statements about collections of quarks, gluons and electrons. The critical question remains, however, How different is more? Is it simply that the novel properties manifested by complex entities require an extended range of concepts for their effective description, or is it the case that emergence is even more interesting than that, in that what is involved requires an enhanced understanding of the causal variety of the world?

The point at issue can be illustrated by considering the wetness of water. This is not a property possessed by a single

3. R. Dawkins, *The Selfish Gene*, Oxford University Press, 1976.
4. F. Crick, *The Astonishing Hypothesis*, Simon and Schuster, 1994.

specimen of H_2O, but it is an emergent effect due to the re-adjustments of the distribution of energy brought about by bringing together a very large aggregation of water molecules. Rather than attempting the impossible task of calculating the mutual interactions of, say, more than 10^{20} molecules, it is convenient to introduce the notion of surface tension in order to think about the behaviour of drops of water. Yet we have every reason to believe that surface tension is simply the macroscopic expression of the consequences of all those microscopic molecular interactions. No causal property of a novel kind is thought to be at work beyond the cumulative effect of intermolecular forces. This kind of phenomenon can be thought of as being *weak emergence*. Metaphysically it is an unproblematic idea, however difficult it may be to unravel scientifically the detail of particular instances.

Strong emergence would correspond to the case in which a new causal principle becomes active in a complex system, of a distinct kind not encountered at lower levels of complexity. More would then be different in a radical way. An example of strong emergence would be if it is indeed the case that human persons possess the power of free agency and are able to act in the world to bring about their choices in a fashion that is not simply an immensely complicated addition of the causal properties of the elementary particles that make up their bodies.

(The question of what degree of freedom humans enjoy in the exercise of agency has, of course, been a matter of long-standing philosophical controversy. Strong emergence would correspond to the so-called liberty of indifference, in which a person makes a choice between genuinely open possibilities, rather than to the liberty of spontaneity, in which actions accord with wishes, but both deed and desire could together be subject to an all-encompassing determination arising from the lowest physical level. The point at issue concerns the status of what philosophers call an incompatibilist account of human freedom, the claim that true personal liberty cannot be reconciled with total physical determinism in the behaviour of constituents. I must confess to being an incompatibilist.)

Is strong emergence of this irreducible kind a conceivable possibility, given what science can tell us about the causal nexus of the world? If the scientific account presented us with a single causal web of known and determinating character, smoothly interpolating between the behaviour of entities encountered at all levels of scientific enquiry but deriving solely from the properties of basic constituents, then the answer would seem to be No. Causality certainly looks like a zero-sum game, and if the causes operating at one level are totally adequate to determine all that happens, the reduction to that level, though doubtless practically infeasible, would surely be ontologically correct. (That is why I am an incompatibilist.) However, the patchwork character of scientific understanding implies that it is by no means certain that such a seamless web of basic causality is the right way to think about science's account of the process of the world. The point can be illustrated from within physics itself before going on to consider the matter in greater generality.

There are two broad types of physical theory employed with great empirical success. One is classical physics, founded on the ideas of Isaac Newton and only somewhat modified in their character by Albert Einstein's great discoveries of special and general relativity. Its picture of the physical world is clear and deterministic. The mathematical expression of classical physics is in terms of differential equations that specify precisely how directly observable physical quantities, such as position and momentum, will vary with time. According to Newtonian thinking, these quantities are believed to be measurable to arbitrary degrees of accuracy and, in principle, they would be completely determined for all time once their initial values were known exactly. Such variables may constitute a discrete set, as in the case of individual particles, or they may carry continuous labels, as in the case of fields spread out through space. The mathematical difference between these two classes simply corresponds to whether the relevant equations are ordinary or partial differential equations. Either type of equation yields a unique solution given a precise and well-posed set of initial conditions. This mathematical property was the source of the conviction, widely held for more than two centuries following the publication of Newton's *Principia*, that the physical world is mechanical, that is to say, tame, controllable and reliably predictable in its behaviour. We shall see that the actual story is more interesting and problematic than that.

The other type of physical theory is quantum physics, brought to a fully developed expression through the great discoveries of the mid-1920s.[5] Its character is quite different from that of classical physics. For one thing, Heisenberg's uncertainty principle asserts that it is impossible to know with arbitrary accuracy both the position and the momentum of a particle. The clear knowledge of initial circumstances assumed by classical physics is in fact unattainable, for the physical world described by quantum theory is intrinsically fuzzy. The scale

5. For an account of quantum ideas, see J. C. Polkinghorne, *Quantum Theory: A Very Short Introduction*, Oxford University Press, 2002.

of irreducible uncertainty is set by Planck's constant (denoted by $\hbar$), a fundamental constant of nature whose value is extremely small in terms of everyday magnitudes. This smallness of $\hbar$ is the reason why quantum physics remained undetected until physicists began to probe phenomena at the level of atomic size or smaller.

People commonly say that quantum theory is different from classical physics because it is indeterministic and it deals with probabilities rather than with certainties. The second half of this statement is true (in most cases quantum physics cannot offer precise predictions), but the first half is not necessarily correct. Probabilities can arise in physics for two quite distinct reasons. One is intrinsic indeterminism; the other is ignorance of all the relevant detail of circumstances. The paradigm case of the latter is the fall of a die, where uncertainty of outcome arises from its dependence on the unknown fine detail of the shaking process, rather than from the caprice of absolute chance. It turns out that the probabilistic character of quantum physics could be interpreted as originating in either of these two ways. The great majority of physicists follow Niels Bohr in considering quantum probability to be an intrinsic property, so that Heisenberg's uncertainty principle is understood to be a principle of real indeterminacy. Yet there is an alternative interpretation, due to David Bohm,[6] in which the underlying dynamics is totally deterministic but the actual outcome depends upon certain factors (called 'hidden variables') to which it is impossible for the experimentalists to gain precise epistemic access. According to this view, the uncertainty principle is simply a principle of unavoidable igno-

6. See D. Bohm and B. J. Hiley, *The Undivided Universe*, Routledge, 1993.

rance; its significance is epistemological rather than ontological. Bohm's theory has only minority support in the physics community, but the choice between him and Bohr cannot be made on empirical grounds, since the two interpretations give identical experimental predictions. The decision, therefore, has to be made for reasons lying outside of empirical science itself. The majority are not simply following the tradition of the tribe in making their choice, but they appeal to considerations such as economy, elegance and lack of contrivance as being features in support of their judgement. The role of criteria of this kind in reaching a decision shows very clearly that questions relating to causality cannot be settled on strictly scientific grounds alone, but they call for acts of metaphysical assessment. This is a theme to which we shall return.

The pioneers of quantum theory immediately faced the problem of how this new physics was related to the classical physics which had proved splendidly successful in so many experimentally sifted domains. Two rather different lines of approach were followed. One was taken by Niels Bohr. He had made very significant contributions to the pioneering but piecemeal developments of early quantum thinking but, by the time a fully articulated theory had come to light through the discoveries of Werner Heisenberg and Erwin Schroedinger, Bohr had moved on to the grandfatherly role of being the philosophical guru of the new physics. He emphasised that physical knowledge comes to us from experimental measurement, and all measuring apparatus is classical in its behaviour. Laboratories are places where definite events happen — the pointer moves to a specific point on the scale and that is that. For Bohr, therefore, quantum theory was about what happens in a world containing these pieces of classical measuring apparatus. Its concern was with what could be said about experiments with electrons, not with what electrons actually were in themselves. He once wrote to a friend that

> There is no quantum world. There is only abstract quantum physical description. It is wrong to think that the task of physics is to find out how nature is. Physics is concerned with what we can say about nature.[7]

One has to say that this positivist notion, asserting that science is simply concerned with talking about the results of experimental measurements, is not at all appealing to most of those who devote their lives to the hard work of scientific research. Rather, they believe that the payoff for all that demanding labour lies in the conviction that they are thereby learning more about the actual nature of the physical world. Without such a realist understanding, the work of science (and the discussions in this chapter) would lose much of their point.

There is also another problem with Bohr's point of view. He seemed to divide the world up into two parts: the accessible realm of classical measuring apparatus and the inaccessible realm of quantum entities, while insisting that they should always be considered linked together in a mutual engagement that he called a 'phenomenon'. This quasi-dualist picture cannot be right. Those reliable instruments in the laboratory are themselves composed of quantum constituents. There are not two distinct domains, but ultimately there is only one physical world.

An alternative way of thinking was to suppose that sys-

7. Quoted in M. Jammer, *The Philosophy of Quantum Theory*, Wiley, 1974, p. 204.

tems become more and more classical in their behaviour as they become 'large', so that a fuzzy world becomes clearer, and probabilities tend to move closer to certainties, as the relevant scale of events grows bigger. The existence of Planck's constant defines a measure in terms of which this idea of largeness can be expressed. It is worth exploring a mathematical way of formulating the notion, even if the going may be a little hard at times for the non-mathematical reader. The attempted trick lies in saying that quantum theory should turn into classical physics in the limit when $\hbar$ tends to zero. Of course, this is a formal device, since actually $\hbar$ is a constant with a specific magnitude in terms of physical units. It cannot literally be set equal to zero. Rather, using this mathematical strategy is a way of trying to express the expectation that classical behaviour will be manifested when Planck's constant is *very small* compared with the magnitudes of the quantities describing the actual system under consideration. It turns out that in some circumstances sense can be made of this idea, and when this is the case the emergence of classical-like behaviour is indeed obtained. Invoking this kind of transition from quantum to classical behaviour is called 'the correspondence principle' and Bohr, in fact, had made clever use of it in his early explorations of quantum phenomena. However, things are not always so straightforward. This is manifested mathematically by the fact that in the general case one cannot just take the formalism of quantum mechanics and put $\hbar=0$. The move fails because the limit of $\hbar$ tending to zero is what the mathematicians call singular. Expressions blow up and formulae become nonsensical. The general relationship between classical physics and quantum physics turns out to be delicate and subtle. This is the main point arising from the discussion that the non-mathematical reader needs to hold onto.

(Technical note: One can get some notion of the subtleties involved by considering Schroedinger's wavelike formulation of quantum physics. Adding two quantum waves, each of intensity 1 and travelling in opposite directions, does not give a single wave of intensity 2, but a wave whose intensity oscillates, varying between 0 and 4. In the limit as $\hbar$ tends to zero, this sum is found to fluctuate with ever-increasing frequency [this is a manifestation of the singularity of the limit]. Only when an average is taken over these rapid oscillations will the intuitively expected classical answer of 2 be obtained. [Think of two candles being twice as bright as one.] This averaging process is required to smooth out the singular behaviour.)

Bohr reminded us that measurement is some kind of engagement between quantum entities and classical-like apparatus. If we are not content to remain at a positivistic distance from what is involved, we shall soon encounter one of the great unsolved mysteries in the interpretation of quantum theory, the 'measurement problem'.[8] Once again our concern is with something for which it is worth a little effort to try to understand the nature of the issue.

A deep way of understanding why quantum theory is radically different from classical physics lies in the counterintuitive principle that the former allows one to add together, in a well-defined sense, states that classical physics and common sense would say are totally immiscible. For example (and the reader will just have to accept this strange fact), according to conventional quantum thinking, an entity such as an electron

8. Polkinghorne, *Quantum Theory*, pp. 44–56.

17

can be in a state which is a mixture of 'being here' and 'being there'. This fundamental feature of quantum thinking is called 'the superposition principle'.

When a measurement of position is made on this unpicturable state, there is a certain probability that the electron will be found in one place and a certain probability that it will be found in the other. Of course, when a definite measurement is actually made (essentially a kind of classical-like intervention by the measuring apparatus on the quantum entity), on any particular occasion one will get a definite answer, though not always the same answer on each occasion that the experiment is repeated. Sometimes the result will be 'here' and sometimes it will be 'there'. This is the point at which probabilities come in. The relative frequencies of these results is something that the theory enables us to calculate with impressive accuracy, but it does not explain how it comes about that a particular answer is obtained on a particular occasion. This latter issue remains unresolved even after almost eighty years of successful exploitation of quantum theory itself. There is no universally accepted way of explaining how the fitful quantum world and the reliable classical world are joined to each other by the bridge of measurement. Here is a level transition within physics that so far remains beyond our understanding. We know how to do the sums, but we do not fully understand how it all works.

Many answers have been proposed to the measurement problem. In brief summary they include: (i) inducing a definite result is an effect of the irreversible behaviour of the large and complex systems that make the measurement (this is often called the (neo-)Copenhagen interpretation); (ii) some speculative form of (so far undiscovered) new physics brings it about; (iii) the intervention of human consciousness induces a determinate result; (iv) everything that might happen actually happens, but with different results appearing in the differing worlds of a proliferating multiverse; (v) it all depends upon the values of hidden variables of a Bohmian type; (vi) quantum physics refers to statistical ensembles and not to individual occurrences, so it just happens and that's that. Even so brief a catalogue makes it plain how diverse are the options that have been canvassed. It is not necessary to go into their details here, but we should note that none of them is wholly satisfactory in its current formulation. In consequence a very important joint in the causal nexus of the physical world remains problematic and controversial. Given the importance of measurement in science, it is embarrassing for a quantum physicist to have to admit this gap in our understanding.

As a result, no smooth and well-understood transition between quantum uncertainty and classical reliability is known to us. In particular, just appealing to a simple division between 'small' and 'large', however intuitively attractive that might at first sight seem to be, does not work. Not only are there many veiled consequences of quantum physics that are fundamental to the existence of a macroscopic world of the kind that is familiar to us (the stability of atoms and molecules would be a simple example), but there are further issues, soon to be considered, that indicate the subtle character of the mutual interpenetration of quantum theory and classical physics. The border between them is fractal-like rather than linear. They intermingle in complex ways. This lack of a fully integrated account of the two physical regimes means that the notion of a 'quantum event' is a much more problematic concept than many of those who appeal to it are prone to recognise.

Another failure of successful integration relates to the two greatest physical discoveries of the twentieth century, quantum theory and general relativity (the modern theory of gravity and spacetime structure). For more than eighty years they have remained imperfectly reconciled with each other, though there are contemporary hopes that the speculations of the string theorists might prove to be an important step in the direction of unification.[9] Presumably when a full theory of quantum gravity is attained, it will radically modify our ideas of the nature of space and time on the smallest conceivable scales (estimated to be about 10^{-33}cm and 10^{-43} sec respectively), since they will become subject to quantum fuzziness and the physical world may be expected to dissolve into some sort of spacetime 'foam' at its lowest level.

Within classical physics itself, there are also subtleties to be considered. The intricate nature of its causal properties is well illustrated by the discovery of chaos theory.[10] Towards the end of the nineteenth century the great French mathematician Henri Poincaré had come to realise that Newtonian theory was not as unproblematically predictable as his distinguished predecessor Pierre Simon Laplace had thought it natural to suppose. The latter had made his celebrated assertion that a 'calculating demon' of unlimited computational power, furnished with the details of the states of motion of all the particles in the universe as they are at the present moment, could predict the whole of the future and retrodict the whole of the past. However, Poincaré realised that there are many classical systems with the property that, though their equations

9. See B. Greene, *The Elegant Universe*, Jonathan Cape, 1999.
10. See J. Gleick, *Chaos*, Heinemann, 1988.

appear mathematically deterministic, the solutions are exquisitely sensitive to the most minute detail of the initial conditions imposed upon them. The slightest variation in these conditions totally changes the expected future behaviour. (A toy model of such sensitivity would be a bead at the top of a smooth wire, bent in the shape of an inverted U. The slightest nudge will cause it to fall, but which side depends upon the fine detail of the disturbance. In more complex cases, bifurcations of possibility of this kind can accumulate without limit.)

The dead hand of mechanism was thereby released from macroscopic process, for the classical world was no longer perceived, even in principle, to be tame and reliably predictable. Putting it more picturesquely, there are at least as many disorderly clouds in the world as there are orderly clocks. Chaotic systems are inseparably linked to their environment, since the slightest external disturbance will radically affect their behaviour.

Systems with this property of extreme sensitivity do not need to be very complicated. An example that Poincaré studied was the problem of three bodies in mutual Newtonian gravitational attraction. There is no general solution of the kind the mathematicians call 'analytic' (smooth, well-behaved), because so many of the possible motions are chaotically sensitive and consequently unstably complicated.

These results of great sensitivity were rediscovered in the era of computerised calculations when Ed Lorenz came to realise that the equations that he was studying as a highly simplified model of a weather system did not produce similar solutions under very small changes of the input data. Instead, tiny initial variations induced completely different subsequent behaviour. The modern theory of chaos stems from this discovery. A number of the properties of chaotic dynamics are of great interest. We shall begin by considering the further subtle complications that chaos theory introduces into thinking about the relationship between classical physics and quantum theory.

The behaviour of a typical chaotic system soon comes to depend upon details of its initial configuration that require a degree of precise specification forbidden by Heisenberg's uncertainty principle. At first sight this fact might seem to offer the prospect of fusing quantum and classical characteristics, a process by which the consequences of the widely supposed indeterminacy of quantum phenomena might be amplified through chaotic dependence upon small detail to produce a widespread causal openness also in macroscopic phenomena. However, this idea is much more problematic than such a sketchy account indicates. The difficulty is that quantum theory and chaos theory do not fit neatly together. The reason for this is somewhat technical, but it is worth pursuing. It centres on the incompatibility of theories that have an intrinsic scale (like quantum theory) and those which do not (like chaos theory).

One way of indicating the mismatch is to note that quantum mechanics characteristically describes time dependence in terms of a discrete set of different frequencies (in this respect the spectra of atoms are similar to the fundamental and harmonics of an organ pipe), while chaotic dynamics has no characteristic frequencies, its time dependence corresponding to what is called mathematically a continuous spectral representation. As a result, quantum systems display periodic properties over long time intervals, a property that chaotic systems precisely do not possess. They are intrinsically aperiodic.

Another (and perhaps for the general reader, an even more mysterious) way of making the comparison is in terms of the differing properties of the geometrical structures that the theories generate in what the physicists call phase space. (One can think of this simply as a useful mathematical device for representing sets of possible motions.) We have already noted the presence in quantum theory of a scale set by Planck's constant. Effectively this means that quantum phase space is coarse-grained into a set of fuzzy regions of a size given by $\hbar$. In contrast, the phase space geometry associated with chaotic systems is fractal in its character, that is to say, it appears essentially the same on whatever scale it is sampled. (The paradigm example of fractal geometry is the celebrated Mandelbrot set. Take a small part of any representation of it, blow up the size of that part, and it will be found to be similar to the whole from which it was taken. The Mandelbrot set is, so to speak, 'the same all the way down' into its infinitesimal depths. It has no natural scale.) The mismatch between behaviour with a scale and behaviour that is scale-free has frustrated the development of a consistent unification of quantum and chaotic thinking in order to produce a theory of quantum chaology. In fact, one expects that the scaling imposed by quantum physics will have the effect of suppressing chaotic behaviour when the latter becomes sensitive to effects at the quantum level. Yet there is more still to be said about the problem, for further investigation of particular cases shows that the plot can thicken even more.

The highly complex character of physical causality in a realistic situation is illustrated by the tale of Hyperion, one of the moons of Saturn.[11] This irregularly shaped lump of rock, about the size of New York City, is observed to be tumbling chaotically as it encircles its planet. A simple estimate of the effectiveness of quantum physics in suppressing chaos, even when applied to a large object like Hyperion, leads to the conclusion that chaotic motion should not last for more than a finite period, in this case about thirty-seven years. While Hyperion has not been under observation for quite so long a time, no one expects that it is shortly about to return to orderly behaviour. It turns out that what will prevent this happening is a further influence, environmental in its character, which must also be taken into account. The effect is called decoherence, and in its turn it suppresses typical quantum effects. Decoherence arises from the fact that Hyperion is bathed in a sea of low-frequency radiation, partly coming from the Sun and partly from the universal cosmic background radiation that fills the universe. Interaction with this radiation represents a kind of repetitive 'measurement' process, continually erasing quantum fuzziness and restoring a classical-like situation, thereby preventing the quantum suppression of chaos from tightening its grip. It is rather like someone trying to get to sleep who continually asks himself 'Am I asleep yet?', thereby keeping himself awake. The decoherent suppression of quantum suppression will keep Hyperion tumbling for a very long time to come.

Until recently almost all thinking in mathematical physics

11. M. Berry in R. J. Russell, P. Clayton, K. Wegter-McNelly and J. C. Polkinghorne (eds), *Quantum Mechanics*, Vatican Observatory/CTNS, 2001, pp. 45–8.

made use of Newton's great discovery of the calculus, a technique perfectly adapted to the discussion of smoothly varying change. The fractal character of chaos is something altogether more jagged. States of motion that at present are infinitesimally close to each other will subsequently separate exponentially far apart. The onset of chaos is often signalled by a cascade of bifurcating possibilities as more and more options open up for future behaviour. In these endlessly diversifying kinds of circumstance, generalisations are called for that go beyond the 'nice' smooth properties that mathematicians associate with what they call 'integrable functions'. It becomes necessary mathematically to consider possibilities associated with jagged 'non-integrable' functions. From a physical point of view these complications arise from the existence of instabilities induced by what are called Poincaré resonances, endlessly active couplings between different possibilities of motion which frustrate the attainment of a clear, separable description. The resulting interlocking complexity dissolves the possibility of an itemised description of the system in terms of distinct trajectories and enforces the need for a purely statistical account of the character of nonintegrable systems. Ilya Prigogine has particularly emphasised this point of view. Speaking of the difference between individualised description (whether in terms of classical trajectories or quantum wave functions) and description in terms of a statistical ensemble, he says, 'Remarkably, at all levels, instability and nonintegrability break the equivalence of both descriptions'.[12] Prigogine believes that the forced move to an ensemble account is the reason for the existence of irreversible processes,

12. I. Prigogine, *The End of Certainty*, The Free Press, 1996, pp. 107–8.

characterised by an increase in entropy (the measure of the disorder in a system).

It has certainly been a long-standing puzzle in physics how fully to understand irreversibility, the emergence in the behaviour of complex systems of a definite direction for the arrow of time, distinguishing the past from the future, despite the fact that these systems arise from the combination of constituents whose fundamental interactions are symmetrical in character between past and future. We know that a film in which a broken glass reassembles itself is a film being run backwards (the true arrow of time of our macroscopic experience always points from order to disorder), but a film of two electrons interacting (were Heisenberg to allow one to record it) would make equal physical sense run in either direction. The reason that the time-reversed event regenerating the broken glass is not feasible is not because it is absolutely forbidden, but because it would require exquisitely precise correlations of the detailed motions and interactions of the returning fragments to a degree that it is just not possible to achieve. Disorder (broken glass) wins out over order (a perfect goblet) because there are overwhelmingly more ways of being disorderly than of being orderly. The increase of entropy in isolated systems, decreed by the second law of thermodynamics, is a manifestation of this preponderant tendency to anarchy. (An everyday analogy is the way that papers pile up higgledy-piggledy on your desk if you do not intervene to tidy it up occasionally.) Many believe that there are connections here with the quantum problem of measurement. Certainly the latter is irreversible, for it defines an arrow of time. Before one was ignorant; afterwards one knows the result. If it is indeed the irreversibility of macroscopic systems that elicits a specific result from making a measurement on a microscopic quantum entity, as the neo-Copenhagen interpretation of quantum theory suggests, this would then provide a striking example of the genuine flow of causal influence from the large to the small (what is often called 'top-down causality').

Yet it would be wrong to interpret the role of disorder as implying that physics describes a domain of the Miltonian 'reign of Chaos and Old Night'. In actual fact, the interlacing of order and disorder is precisely what seems to be needed for the creative emergence of novelty. New things happen in regimes that we have learned to identify as being 'at the edge of chaos'. Too far on the orderly side of that frontier and things are too rigid for there to be more than a shuffling rearrangement of already existing entities. Too far on the disorderly side, and things are too haphazard for any novelties to persist. A simple example of this principle is afforded by biological evolution. Without a degree of genetic mutation, life would be frozen into the existing range of forms. Too high a mutation rate, and there would be no quasi-stable species on which natural selection could operate.

'Chaos theory' has turned out to be an unfortunate misnomer. Order and disorder are found to interlace within its scenarios. All is not radical randomness. In dissipative chaotic systems (those in which friction operates), behaviour soon converges onto an intricate but limited portfolio of possible forms, called a 'strange attractor'. ('Attractor' indicates that motions converge upon it; 'strange' refers to the fractal character of its structure in phase space.) In those cases where chaos is generated through a cascade of bifurcating possibilities, there is a remarkable universal pattern in the way this happens, characterised by a new mathematical constant of fundamental significance, discovered by Michael Feigenbaum.[13]

A particularly interesting class of systems at the edge of chaos are those dissipative physical systems that are held far from thermal equilibrium by the continuous exchange of energy and entropy with their environment. All living entities have this character. Complete thermal equilibrium is a dull state of maximum entropy in which there is nothing really interesting left to happen. In contrast, systems far from equilibrium are ones in which small fluctuations can trigger the spontaneous generation of astonishing patterns of large-scale dynamical behaviour.[14] (Because these systems are not isolated but they are coupled to their environment, the second law of thermodynamics does not forbid this generation of orderly pattern, since a compensating degree of disorder is exported into the surroundings.) A simple example of this behaviour is provided by Bénard convection. When a fluid contained between two horizontal plates is heated from below, an appropriate temperature difference between the bottom and top can induce a striking phenomenon in which the convective motion of the fluid is contained within a pattern of hexagonal convection cells. This behaviour corresponds to the correlated motion of trillions upon trillions of molecules.

The unexpected degree of novel behaviour displayed by dissipative systems certainly illustrates the point that that more can be different. Similar properties have been discovered by complexity theorists, whose work so far has mostly centred on the natural history–like observation of the properties of

13. See Gleick, *Chaos*, pp. 171–81.
14. I. Prigogine and I. Stengers, *Order out of Chaos*, Heinemann, 1984.

moderately complicated computer-generated models. Stuart Kauffman has discussed a particularly interesting example.[15] Its logical form is that of a Boolean net of connectivity 2, but it is easier to visualise the system in terms of a hardware equivalent. Consider a large array of electric light bulbs, each of which has two possible states, 'on' and 'off'. The system develops in steps and there are simple rules determining its state at the next step from its state now. Each bulb is correlated with two other bulbs somewhere else in the array and the rules specify how the bulb's next state is derived from the present state of the two bulbs that are its correlates. The array is started off in a configuration of random illumination, some bulbs on and some bulbs off. It is then left to develop according to the rules. One might have expected that generally nothing very interesting would happen and that the array would just twinkle away haphazardly for as long as it was let to do so. In fact, this is not at all the case. If there are 10,000 bulbs in the array, the number of possible states of illumination possible in principle is $2^{10,000}$, or about $10^{3,000}$, an absolutely huge number. Yet, the system soon settles down to cycling through only about one hundred states! (In more general terms, if there are N bulbs in the array, the number of patterns of illumination finally realised in this way is about $N^{1/2}$.) This phenomenon represents the spontaneous generation of an altogether astonishing degree of order. Once again, more seems to be different in a highly non-trivial manner. Similar spontaneous generation of complex order is also found to occur in systems of cellular automata.[16]

15. S. Kauffman, *At Home in the Universe*, Oxford University Press, 1995, ch. 4.

16. S. Wolfram, *A New Kind of Science*, Wolfram Media, 2002.

At present no general theory is known that covers the behaviour of complex systems, either physical or logical, though the remarkable results sporadically discovered through the investigation of particular cases strongly encourage the expectation that there is a deep theory underlying these examples and awaiting discovery. Progress towards a general understanding may be expected to require a revolution in scientific thinking in the twenty-first century at least as great as that wrought by the discovery of quantum theory in the twentieth century. Two features may be expected to characterise this conceptual development when it arrives.

One is the recognition of the inadequacy of a merely reductionist account, so that the addition of a complementary holistic discourse will be needed, treating entities in the integrity of their wholeness. More is indeed different. None of the remarkable pattern-generating properties of the systems discussed could have been guessed from thinking about their individual components.

It is interesting to note that even so superficially reductionist a subject as the quantum physics of elementary particles provides support for the need for an holistic dimension in our approach to physical reality. Albert Einstein, Nathan Rosen and Boris Podolsky discovered in 1935 that quantum theory implied a counterintuitive togetherness-in-separation (non-locality) for two quantum entities that had interacted with each other. If they separate from each other, even to a great distance apart, they nevertheless remain mutually entangled with each other, so that acting on the one 'here' (say, measuring one of its properties) has an immediate causal effect on the other, no matter how far away it may be. (This is the celebrated 'EPR effect'.)[17] What happens to the distant entity depends specifically upon what happens to the near one, so that the effect is genuinely causal and not merely epistemological, as if it just involved learning about something that was already the case. Of course, there is nothing surprising about the latter. If there are two balls in an urn, one white and the other black, and we each take out one in our closed fists, then if you go a mile down the road before I open my hand, if I see a white ball I know immediately that you have a black one. This was always so and all that has happened is that I have now learned that it is the case. In contrast, the EPR effect is rather like saying that if I found a red ball, then your ball would have had to be to be green, but if I had found a blue ball, yours would have had to be orange. Einstein himself felt that the EPR phenomenon was so 'spooky' that its prediction showed there must be some shortcoming in quantum theory, but later experiments have confirmed that non-locality is indeed a property of nature. Even subatomic particles, it seems, cannot properly be treated atomistically! The apparent localisability of the objects of our everyday experience is not as unproblematic as it might have seemed to be, a thought reinforced when one remembers the sensitivity of chaotic systems to the smallest disturbance arising from their environment.

The other conceptual development will surely be the placing of 'information' alongside 'energy' to form a joint basis for fundamental thinking. By information is meant something like the appropriate specification of dynamical patterns of ordered behaviour. While the concept of information is intuitively appealing, and the idea of it has been cited

17. Polkinghorne, *Quantum Theory*, ch. 5.

recently in many contexts of reflection on the character of the natural world, its specification in more precise terms is a matter of continuing discussion and difficulty. Two possible sources of insight, in particular, have been canvassed. One has been Claude Shannon's theory of communication; the other is computer science.[18] In the former case an important aim is to prevent messages becoming garbled because of interference from background noise. In order to combat this degradation, a degree of redundancy is desirable in the form of the message. (Most mail addresses include both the street name and the postcode. If the postman cannot decipher the one, he may fare better with the other and so the letter will get through.) The appeal to computer science has been aimed particularly at finding some acceptable measure of intrinsic complexity. An informational statement written in binary code (os and 1s) might be very long but the programme specifying it might be very much shorter, for example 'print 01 a million times'. In this case the specification is said to be algorithmically compressible — it is much shorter than the message itself — and the system it refers to is correspondingly simple. In the converse case, algorithmic incompressibility becomes a signal of complexity and potential richness of informational specificity. However, there are no general meta-algorithms for determining the shortest programme that will do the trick.[19] Difficulties of this kind beset approaches to information that rely on computer science. In addition, like the ideas derived from communication theory, they are strictly syntactical in

18. Gleick, *Chaos*, pp. 255–9, II. C. von Bacyer, *Information*, Weidenfeld and Nicholson, 2003.

19. See D. Ruelle, *Chance and Chaos*, Princeton University Press, 1991, chs 21–23.

character and so they bracket out any consideration of the semantic issues of meaningfulness and significance.

We have seen that there is much evidence to support the thesis that 'More is different', since strikingly novel properties emerge in complex systems. There remains, however, the central question of the kind of emergence that is being observed. Is it only weak or can we suppose there are cases where the strong criterion of causal novelty is satisfied? In terms of a hierarchy of levels of complexity, notions of 'top-down causation', by means of which the whole influences the behaviour of the parts,[20] or of the 'supervenience' of a higher level upon a lower, so that both kinds of discourse are simultaneously admitted,[21] are certainly attractive, but they are not unproblematic and in the absence of an actual causal analysis they would seem to be aspirations rather than proposals. We shall not really be able to speak of strong emergence unless it can be argued that the lower level does not soak up all the available causal room for manoeuvre.

We have seen that the nature of the causal nexus of the world is ultimately a matter for metaphysics rather than physics. If the latter reports an epistemic deficit in its account due to the existence of intrinsic unpredictabilities (as both quantum theory and chaos theory actually indicate), then there is an opportunity for the metaphysician to propose that these represent not merely ignorances arising from unfortunate epistemological limitation, but they correspond to actual ontological openness, allowing the operation of further causal principles, active in bringing about the future, beyond those

20. Peacocke, *Theology*, pp. 53–5, 157–60.
21. N. Murphy in R. J. Russell, N. Murphy, T. C. Myerling and M. A. Arbib (eds), *Neuroscience and the Person*, Vatican Observatory, 1999, pp. 147–64.

that a reductionist physical theory is able to describe. Those who embrace a realist philosophy of science, aligning epistemology and ontology as closely as possible to each other from the conviction that what we know or cannot know is a reliable guide to what is the case, would seem almost to be obligated to take this point of view.

The 'laws' of physics discovered at low levels of complexity would then simply be 'downward-emergent' approximations to the character of a more subtle and supple causal story in which the whole truly did influence the behaviour of the parts. (I have called this latter stance 'contextualism'.)[22] The approximation involved in this downward emergence would correspond to the idealisation that the system considered could effectively be treated as being isolated from the influence of its total environment. Such circumstances are sometimes attainable—the feasibility of experimental science depends upon this being the case, for otherwise it would not be possible to investigate anything without having to take everything into account—but the exquisite sensitivity of chaotic systems to any disturbance arising from their surroundings, together with the EPR effect, shows that isolatability is far from being a universal property. We therefore have no compelling grounds for regarding current theories as being more than a form of approximation to actual physical reality as it is encountered in the limit of effective isolatability. While the laws of classical physics taken at face value do indeed exhibit the phenomenon of 'deterministic chaos' (apparently haphazard behaviour arising from solutions of mathematically pre-

22. J. C. Polkinghorne, *Science and Christian Belief/The Faith of a Physicist*, SPCK/Fortress, 1994/1996, p. 29.

cise and deterministic equations), it is a matter for *metaphysical* decision, similar to the decision between the rival interpretations of quantum physics offered by Bohr and Bohm, to conclude what this tells us about the causal nexus of the world. I personally have espoused a realist interpretation that treats chaotic unpredictability as a sign of ontological openness.[23] The expectation of a generalised correspondence principle linking the true theory to its localised approximation encourages the thought that appeal to significant general features of chaos (sensitivity to detailed circumstance; structures similar to strange attractors) will not prove misleading, even though the two accounts must differ in certain respects, as the discussion of the problems of quantum chaology has already made clear.

Making use of science's account of future behaviour in this open metaphysical way by no means demands us to abandon the principle of sufficient reason, requiring a full explanation of the origin of what actually occurs. It is simply to conceive that the portfolio of causes that bring about the future is not limited solely to the description offered by a methodologically reductionist physics and framed only in terms of the exchange of energy between constituents. Instead, the concept of causal influence can be broadened at least to include holistic effects of an informational, pattern-forming kind. One might call this top-down form of causality 'active information'. Such a move could then represent a small step in the direction of the more ambitious metaphysical programme represented by *dual-aspect monism*. This philosophical scheme aims to treat

23. J. C. Polkinghorne, *Belief in God in an Age of Science*, Yale University Press, 1998, ch. 3; *Faith, Science and Understanding*, Yale University Press, 2000, pp. 99–101.

the mental and the material as complementary poles of the one world-stuff, perceived in different modes of encounter.[24] Regarded from this wider metaphysical standpoint, the duality of energy and information that science is beginning to embrace might prove to be part of a movement towards the attainment of an understanding that takes with equal seriousness our basic human experiences of physical embodiment and of personal agency. The task of carrying this project further would, of course, require very much greater enrichment of the concept of information, moving it on from purely syntactical considerations to enable it to accommodate also semantic dimensions of critical significance in relation to an account of intentional action. Not only would this move provide an improved basis for the understanding of human agency, so that the metascientific discourse would begin to describe a world of which we could fittingly conceive ourselves as being inhabitants, but it could also refer to a physical world within whose open grain it would be fully conceivable that the God who is that world's Creator is providentially at work through the input of active information into its unfolding history, in a manner that operates non-interventionally within the grain of nature, rather than interventionally against it. This is an idea that I have explored elsewhere.[25]

24. T. Nagel, *The View from Nowhere*, Oxford University Press, 1986, pp. 28–32.

25. Polkinghorne, *Belief in God*, ch. 3. For a defence of the idea that God may act causally within the openness of the created order, rather than in some ineffable way unique to deity, see J. C. Polkinghorne (ed.), *The Work of Love*, SPCK/Eerdmans, 2001, pp. 104–5. For surveys of current ideas about divine action, see P. Clayton, *God and Contemporary Science*, Edinburgh University Press, 1997, chs 5–8; N. Saunders, *Divine Action and Modern Science*, Cambridge University Press, 2002; W. Wildman, 'The Divine Action Project, 1988–2003', *Theology and Science*, 2, pp. 31–75 (2004).

At present, these ideas necessarily remain largely hopes for future understanding. What one can say, both in relation to human agency and in relation to divine providential action, is that the proper acknowledgement of our fragmented knowledge of the causal structure of physical reality is at least sufficient to 'defeat the defeaters', to challenge and put in question those who are trying to assert the necessity of a merely physical reductionist view. It is clear that science has not demonstrated the causal closure of the natural world. Nothing it can tell us requires us to deny our directly experienced human capacity for intentional action, nor can science forbid religious believers to hold to their belief in God's providential interaction with the history of the world.

CHAPTER THREE

Human Nature: The Evolutionary Context

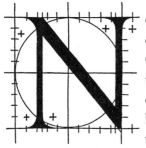o scientist has had greater influence on general human thinking than Charles Darwin. The publication of the *Origin of Species* in 1859 was a climactic point in a process started by the geologists towards the end of the eighteenth century, which recognised the inherent changeableness of the natural world and the consequent necessity of an historical mode of thinking if one is to gain a proper understanding of its character. It is impossible for us today to consider the present without seeking to take into account its origin in the past. Time is no longer simply the index of when events happened, but it signifies and contains the evolutionary process through which things have come to be.

A world in which species were stable, totally immune from change, might well have been capable of being thought of theologically as a creation that had sprung into being ready-made, its origin simply the result of the direct action of the

God who was that world's Designer. On the other hand, a world of radical temporality, in which change is the engine driving the emergence of novelty, is one to which its Creator's relationship has to be understood in somewhat different terms. In words used by both Charles Kingsley and Frederick Temple in the aftermath of the publication of the *Origin*, an evolving world may appropriately be thought of theologically as a creation in which creatures are 'allowed to make themselves'. In other words, from a theological perspective, evolution is simply the way in which creatures are allowed to explore and bring to birth the fruitfulness with which the Creator has endowed creation.

When the initial impact of Darwin's great discovery had somewhat abated, other theologians followed in the wake of Kingsley and Temple in welcoming these new insights. Many came to see that evolutionary thinking was not incompatible with a doctrine of creation, but the latter would have to be expressed somewhat differently from the way that had been customary in the past. Emphasis now needed to be laid not only on *creatio ex nihilo* (understood as asserting the will of God to be the cause not only of how things began but also why things remain in existence), but also on a complementary process of unfolding *creatio continua*. With the benefit of hindsight, this story of development might even be thought, anachronistically, to have been hinted at in the sequence of the 'days' of creation in Genesis 1, and by Augustine's notion that the initiating timeless act of creation brought into being the 'seeds' from which eventually a multiplicity of different creatures would develop in the course of a process of temporal germination. Such ideas are not at all uncongenial to a theology that sees God as the ordainer of the character of creation, and so as the

One who is to be thought of as acting as much through natural processes as by any other means. Unfolding evolution simply expresses the divine intention for the way in which creation is to realise its God-given potentialities. The contingencies present in the process represent the Creator's gift to creatures of the freedom to make themselves. To suppose the contrary, and to posit an opposition between natural process and divine purpose, would be to fall into the Manichean heresy, the supposition that God and the world are totally at odds with each other.

Much more shocking to some theologians, and to much human sensibility generally, was the idea, implicit in the *Origin* but only later stated clearly by Darwin (who was always somewhat apprehensive about public reaction to what he had to say, so that he waited until the publication of *The Descent of Man* in 1871 before he made the point explicitly) that evolutionary thinking must embrace human origin as well as that of the other creatures. The consequent implication of humanity's close degree of kinship with the animals was by no means a welcome thought to all. Many, whether religious believers or not, felt that it carried the threat of an intolerable subversion of human dignity and status. This fear underlies the unedifying exchange about monkey ancestry said to have taken place between Bishop Samuel Wilberforce and 'Darwin's bulldog', Thomas Henry Huxley, in their notorious encounter at the British Association meeting at Oxford in 1860.[1]

It is impossible for us today to think about human nature without acknowledging the significance of the evolution-

---

1. There seems to be some historical uncertainty about what actually happened; see J. H. Brooke, *Science and Religion*, Cambridge University Press, 1991, pp. 40–2.

ary origin of *Homo sapiens*. While this fact certainly implies a degree of cousinly relationship between humankind and the other animals (particularly the higher primates), it by no means implies that this recognition exhausts all that needs to be said, as if human beings are just another kind of ape. The history of the development of life has been characterised by a sequence of unprecedented emergences, leading to the appearance on the terrestrial scene of that which is qualitatively novel. The first such emergence was life itself, when the complexifying organisation of inanimate matter reached a stage that produced systems that could maintain and reproduce themselves. After about two billion years during which all living entities were single-celled, there emerged the increasing complexity of multicellular organisms. Eventually this led to the dawning of consciousness for the first time on planet Earth. In turn, we should not hesitate to recognise that there are a number of characteristics of human nature, and perhaps of the natures of our immediate hominid ancestors, that clearly mark out the genus *Homo* as constituting a yet further emergent level of fundamental and astonishing novelty. Of course, we have evolved from previous forms of animal life, but one should not commit the genetic fallacy of supposing that knowledge of origin is the same as knowledge of nature. I want briefly to survey a number of characteristics that support a claim for unique human status.

The first point is that humans are *self-conscious beings* in a radically new way. Of course, the higher animals are conscious, but they seem to live in what we may call the near present. The chimpanzee can foresee that throwing the stick may dislodge the banana, and understand that the person who just now went behind the rock is still there though hidden from sight, but the power to look far ahead, even to the point of brooding on the thought of our eventual deaths, despite the fact that they may lie many years in the future, seems to be a capacity that only humans possess. It is said that elephants show signs of sorrow at the death of one of their number, but again this is something related only to the near present. Part of humanity's unique self-consciousness is a keen awareness of ourselves, not just as recognised when we see our images reflected in a mirror, but as persons whose characters are formed by our experiences of life, reflectively and reflexively assessed.

The human possession of *language*, with its profound conceptual range and almost limitless flexibility to respond to novel experiences and changing situations, is clearly linked with the human exercise of self-conscious reflection. It enables that very characteristic human activity of story-telling, and the remarkable possibility of writing poetry. As with a number of other human characteristics, one can see some prefiguration of linguistic abilities in the capacities possessed by the higher apes and some cetaceans,[2] but the differences in degree between them and human beings remain so great as to amount to a qualitative distinction. It is certainly interesting that prolonged human intervention can induce in selected chimpanzees the ability to manipulate a limited vocabulary of signs and to form simple 'sentences', but these achievements fall very far short indeed of the creativity of human linguistic attainments.

More generally, human beings possess a great range of *rational skills*. Later we shall pay some attention to mathemat-

2. See the discussion of the communication skills of dolphins in S. Conway-Morris, *Life's Solution*, Cambridge University Press, 2003, pp. 250-3.

ics, but for the moment one may simply illustrate the fruit-fulness of human reasoning by appealing to the astonishing extent of our scientific understanding. With the dawning of hominid self-consciousness not only did the universe become aware of itself, but a process began through which the secrets of its structure and history would progressively come to be unveiled. Even such counterintuitive regimes as the subatomic world of quantum theory, or the vast expanses of cosmic curved spacetime, radically different in their character from the world of everyday experience and remote from direct impact on it, have proved to be open to human enquiry and understanding. Human ethologists write substantial volumes analysing the behaviour of communities of chimpanzees, thereby displaying an insightful interest in their character, but this is not reciprocated by the primates in their turn. Learning how to use a stone to crack a nut is a valuable skill, but it scarcely bears comparison with the remarkable devices produced through human technological invention.

Human beings also possess great *creative powers*, manifested through art and culture. From the time of the earliest cave paintings known to us, there is evidence of an engagement with the aesthetic, displaying a quality that defies explanation in merely pragmatic terms. The beautiful notes of birdsong are apparently principally a means of asserting territorial possession, but humans explore the inexhaustible riches of music for reasons that centre on delight rather than utility.

We are also *moral beings* in a way that does not seem to be the case for the animals, for whom concepts of right and wrong and of ethical obligation do not appear to be appropriate. When we read medieval stories of an ox being put 'on trial' for goring a man to death, and then 'executed' for the misdeed, they seem to us to be grotesque, the result of a foolish kind of category mistake. I am sure that we should treat animals with ethical respect, but grounding this attitude in a doctrine of animal rights seems to me to be mistaken, for the use of rights language would surely require there to be a complementary doctrine of animal duties, a notion that does not appeal to us as being reasonable.[3]

At almost all times and in almost all places, human beings have participated in an admittedly bafflingly diverse history of encounters with the sacred, experiences that indicate that we possess a capacity for what may fittingly be called *God-consciousness*. Mystical apprehension of unity with the One or the All; numinous encounter with the mysterious and fascinating reality of the divine standing over and against humanity in mercy and judgement; the less dramatic engagements of regular worship; all these are human experiences of great intensity which many of us value as being of undeniable authenticity and significance, but which have no discernible counterpart in the lives of our animal cousins.

Finally, theologians detect in human life a slantedness which they categorise as *sin*, a source of distortion in human affairs that frustrates hopes and corrupts intentions. Of course, among the animals there is not only necessary predation, but also what appears to be a kind of cruel mischief, as when apes sometimes wantonly tear apart small monkeys whom they have happened to come upon. Of course, among the animals there can be fierce struggles over access to food or mating, or a challenge to the hegemony of the α-male, though

3. Of course, human neonates have rights but not yet duties, but they will grow into an ethical awareness that will bring with it moral responsibility. We do not see signs of a corresponding ethical development in animals.

these are not often fought literally to the death. In contrast, there appears to be in human history, with its wars, crusades and acts of genocide, a degree of depravity that seems to be on an entirely different scale from these animal incidents. And these communally willed human acts of immense evil are the shadows writ large of the many lesser acts of betrayal and selfishness that occur all the time in the individual circumstances of everyday life. We are not only moral beings, but through our actions we show ourselves to be flawed moral beings. Reinhold Niebuhr once observed that the only empirically verifiable Christian doctrine is that of original sin. Simply look around you, or within your own heart, and you will see that it is true. Theologians diagnose the root of sin as lying in human alienation from the God whose gift of the spiritual power that theologians call grace is the proper ground of a fully human life, a theme to which we shall return.

Even so brief a survey indicates how strange it is that many biologists can claim not to be able to see anything really distinctive about *Homo sapiens*. They regard human behaviour as just another instance of animal behaviour, and humanity as a not particularly special twig on the burgeoning bush of evolutionary development. In fact, even the ability to articulate these assertions is sufficient to deny their premise, providing as it does an example of humankind's unique capacity for self-reflexive thought. The fact that we share 98.4 percent of our DNA with chimpanzees shows the fallacy of genetic reductionism, rather than proving that we are only apes who are slightly different. After all, I share 99.9 percent of my DNA with Johann Sebastian Bach, but that fact carries no implication of a close correspondence between our musical abilities.

I must confess to being a speciesist—provided that term

is understood as involving an acknowledgement of the true novelty resulting from hominid emergence, but not if it is taken as implying a failure to accord the proper kind of respect to the animals. Yet, the ethical basis for this respect takes a different form from that which underlies the conviction that human persons are each of unique value, so that they are never to be treated as means but only as ends. Many of us consider that it is permissible in an humane way to cull a herd of deer facing a severe shortage of fodder, thereby preserving the type at the expense of the deaths of some of the individuals. Such a strategy in relation to a human community faced with famine would universally be acknowledged to be unethical.

The concern of this chapter is to take absolutely seriously the fact that human beings have emerged through a long and continuous history of biological evolution, while at the same time taking with equal seriousness the fact of the qualitative differences that correspond to the irreducible novelty of that emergence. The first consequence of acknowledging human evolutionary origin is to reinforce an understanding of human beings as *psychosomatic unities*. While a dualist picture of humans as possessing a spiritual soul encased in a material body from which it is potentially separable is not absolutely ruled out—for one could conceive of the Creator as bestowing an extra dimension of spiritual being on a purely material entity that had attained the degree of complex development that would make this gift appropriate (something like this appears to be the official Vatican view)—it is surely much more persuasive to think of humans as animated bodies, a kind of 'package deal' of the material and the mental and spiritual in the form of a complementary and inseparable relationship (p. 36). The aim is to take human embodiment absolutely seriously, without falling into the error of treating our mental and spiritual experiences as no more than epiphenomenal fringe effects of the material. Such an integrated understanding of humanity would not have surprised the writers of the Bible, for this was the predominant way in which Hebrew thought regarded human nature.

Taking a psychosomatic view of human beings requires careful thinking about how to understand the nature of the human *soul*.[4] The concept is one of central importance to theology, where the soul has the role of being the carrier of the intrinsic essence of individual personhood—the 'real me', as one might say. Theological anthropology cannot abandon this use of soul language, though it has to be able to free itself from bondage to a past heritage of conceiving it in terms of platonic categories.

The 'real me' is certainly not to be identified merely with the matter of my body, for that is continually changing through the effects of wear and tear, eating and drinking. What maintains continuity in the course of this state of atomic flux is the almost infinitely complex *information-bearing pattern* in which the matter of the body is at any one moment organised (cf. p. 31). It is this pattern that is the human soul. The idea has a venerable history, for both Aristotle and Thomas Aquinas thought of the soul as being the 'form' of the body, its constitutive and organising principle, so to speak—a concept that obviously has a close connection with the position being advocated here. Three things must be said about this modern version of the way of conceiving of the soul.

4. J. C. Polkinghorne, *The God of Hope and the End of the World*, SPCK/Yale University Press, 2002, pp. 103-7.

The first point is that, as I have already acknowledged (p. 36), the flat language used in the phrase 'information-bearing pattern' is an almost wholly inadequate attempt to point to a necessarily much richer concept, lying beyond anything that we are presently able to articulate properly. Human persons are relational beings and the patterns that constitute them cannot simply end at their skin. The rich variety of human capacities and forms of experience, briefly surveyed in the course of the defence of human uniqueness, must also find appropriate incorporation into the nature of the human soul. The concept of information must be enriched sufficiently to accommodate the mental and spiritual dimensions of human nature. Aristotle and Aquinas thought of the soul as being the principle of life animating a body, and they believed that there are vegetative and animal souls as well as human souls. In terms of the present discussion, what differentiates the human soul would be precisely the rich, many-layered complexity that is a reflection of the unique range of human capacities.

The second point is that the picture I am proposing is a dynamic one, for the pattern that is my soul will develop as my character forms and my experiences, understandings and decisions mould the kind of person that I am. John Keats's image, in one of his letters to Fanny Brawne, of this life as a vale of soul-making is an apposite one. Understood in this way, the soul is not a once-for-all gift, as if it were fully conveyed at conception or at birth, but it has its individual history. This observation does not rule out there being an unchanging component in the soul, which could be thought of as the signature identifying a specific person, but this would only be a part of what makes it up. (The individual genome would, presumably, be a constituent of this invariant sub-pattern.)

The third point is to recognise that this way of understanding the soul implies that it does not of itself possess an intrinsic immortality. As far as a purely naturalistic account could take us, the information-bearing pattern carried by the body would be expected to dissolve at death with that body's decay. It is, however, a perfectly coherent possibility to deepen the discussion by adding a theological dimension, and to affirm the belief that the God who is everlastingly faithful will preserve the soul's pattern *post mortem* (holding it in the divine memory is a natural image), with the intention of reconstituting its embodiment in a new environment through the great divine eschatological act of the resurrection of the whole person. This way of formulating the Christian hope of a destiny beyond death, which would have been perfectly acceptable to Aquinas, is one that I have explored elsewhere.[5] Its basic structure is the pattern of death and resurrection, rather than the notion of spiritual survival.

It is important to recognise that evolutionary thinking is as much concerned with environment as it is with genetics. This point has been emphasised particularly by Holmes Rolston in his reflections on biological process.[6] It was Darwin's great genius to understand how interaction between individual variations (only later recognised as stemming from genetic mutations) and the constraint of survival in a specific ecological setting could together, through the shuffling exploration and sifting of small differences, so shape the development of living forms over long periods of time as to induce a fruitful history of increasing complexification.

5. Ibid., 107-12.
6. H. Rolston, *Genes, Genesis and God*, Cambridge University Press, 1998.

Orthodox neo-Darwinian thinking conceives of the environment relevant to evolution, including that of humanity, as being one that can be understood solely in physico-biological terms, and the mode of interaction with it as being solely the process of differential reproductive success. While we can agree that natural selection has been an important factor in the development of life on Earth, it is by no means obvious that it is the only type of process involved.[7] The timescale of the history of terrestrial life is certainly long (3.5 to 4 billion years), but the train of developments that has to be accommodated within that span is immensely complex, with the first two billion years or so being taken up solely by single-celled organisms. Argument on this general point is not, however, our present concern. Instead, our focus is on the emergence of hominid life, taking place over a period of just a few million years and with modern *Homo sapiens* only appearing in the last one to two hundred thousand years. The central issue for this chapter is whether a strictly neo-Darwinian account, with its narrow concept of the effective environment within which these developments took place, is sufficient to explain the coming-to-be of the many distinctive features that we have claimed mark off human nature from other forms of animal life.

In fact, the attempt to force classical Darwinian thinking into the role of an explanatory principle of almost universal

7. For discussion of the role of the self-organising properties of complex systems, see B. Goodwin, *How the Leopard Changed Its Spots*, Weidenfeld and Nicholson, 1994; S. Kauffman, *The Origins of Order*, Oxford University Press, 1993. See also the evidence for the convergence of different genetic lines onto constrained possibilities for biologically accessible and functionally useful structures, discussed in Conway-Morris, *Life's Solution*.

scope[8] has proved singularly unconvincing as it seeks to inflate an assembly of half-truths into a theory of everything. Sober evaluation of the adequacy of the insights being proffered soon pricks this explanatory bubble. Increasing ability to process information coming from the environment is clearly an advantage in the struggle for survival, but this does not explain why it has been accompanied by the property of conscious awareness. Indeed, one might suppose that the latter, with its limited focus of attention and no more than a peripheral openness to signals coming from other possible directions, might be more a hazard than a help.

Evolutionary epistemology[9] has attempted to explain and validate human rational powers to attain reliable knowledge as being something originating through Darwinian development. Once again one encounters a half-truth. Of course, being able to make sense of everyday experience is a vital survival asset. If one could not figure out that stepping off a high cliff was a dangerous thing to do, life would soon be imperilled. Yet when Isaac Newton recognised that the same force that makes the high cliff dangerous is also the force that holds the Moon in its orbit around the Earth and the Earth in its orbit around the Sun, thereby going on to discover universal gravity, something happened that went far beyond anything needed for survival. When Sherlock Holmes and Dr Watson first meet, the great investigator feigns not to know whether the Earth goes round the Sun or the Sun around the Earth.

8. D. Dennett, *Darwin's Dangerous Idea*, Simon and Schuster, 1995; E. O. Wilson, *Consilience*, Knopf, 1998.

9. P. Munz, *Our Knowledge of the Growth of Knowledge*, Routledge & Kegan Paul, 1985; F. Wuketits, *Evolutionary Epistemology*, State University of New York Press, 1990; W. van Huyssteen, *Duet or Dual?*, SCM Press, 1998.

He defends his apparent ignorance simply by asking what does it matter for his daily work as a detective? Of course, it does not matter at all, but human beings know many things that neither bear relation to mundane necessity nor could plausibly be considered simply as spin-offs from the exercise of rational skills developed to cope with those necessities. This point was reinforced about two hundred years after Newton when Albert Einstein's discovery of general relativity produced the modern theory of gravity, capable of explaining not only the behaviour of our little local solar system but also the structure of the whole cosmos. In both relativity theory and quantum theory, modes of thought are required that are totally different from those appropriate to everyday affairs. Yet, as we have already noted, the human mind has proved capable of comprehending the counterintuitive world of subatomic physics and the cosmic realms of curved spacetime.

It has turned out that it is our mathematical abilities that have furnished the key to unlock deep secrets of the physical universe. Once more one encounters a mystery impenetrable to conventional evolutionary thinking. Survival needs would seem to require no more than a little arithmetic, some elementary Euclidean geometry, and the ability to make certain kinds of simple logical association. Whence then comes the human ability to explore non-commutative algebras, prove Fermat's Last Theorem, and discover the Mandelbrot set? These rational feats go far beyond anything susceptible to Darwinian explanation.

The ability to use the experience of yesterday as a guide to coping with the challenges of today is clearly a significant aid to survival. But does this fact alone give us sufficient licence to trust in human ability to reconstruct from fragmentary evidence the history of a past extending over many millions of years? Darwin himself felt some doubts on this score, writing in old age to a friend that 'With me the horrid doubt arises whether the convictions of man's mind, which has been developed from the mind of lower animals, are of any value or are trustworthy'.[10] There is something touching in this spectacle of this great scientist poised with rational saw in hand and tempted to sever the epistemic branch on which he had sat while making his great discoveries. Surely his doubts were unjustified. The cumulative power of scientific thinking has vindicated itself many times over in the course of human investigations into reality. Why science is possible in this deep way is a question which, if pursued, would take us well beyond science itself.[11] I believe that ultimately it is a reflection of the theological fact that human beings are creatures who are made in the image of their Creator (Genesis 1:26–27).

Sociobiology[12] seeks to explain human ethical intuitions in terms of inherited patterns of behaviour favouring the propagation of at least some of an individual's genes. Once again, one may acknowledge a source of partial insight. No doubt ideas of kin altruism (the mutual support extended between those who share in the family gene pool) and reciprocal altruism (favours done in the expectation of favours later to be received) shed some Darwinian light on aspects of human behaviour. Games theoretic models of behavioural strategies

10. Quoted in M. Ruse, *Can a Darwinian Be a Christian?*, Cambridge University Press, 2001, p. 107.
11. See J. C. Polkinghorne, *Beyond Science*, Cambridge University Press, 1996.
12. E. O. Wilson, *Sociobiology*, Belknap Press, 1975.

that optimise probable returns in given circumstances — such as 'tit for tat': respond in the same manner that your opponent has displayed to you — give some insight into the nature of prudent decision making. But sociobiology tells too banal a story to be able to account for radical altruism, the ethical imperative that leads a person to risk their own life in the attempt to save an unknown and unrelated stranger from the danger of death. Love of that incalculable kind eludes Darwinian explanation.

Equally elusive to evolutionary explanation are many human aesthetic experiences. What survival value has Mozart's music given us, however profoundly it enriches our lives in other ways?

The proper response to all this is not to adopt a Procrustean technique of chopping down the range of human experience until it fits into a narrow Darwinian bed, nor is it to abandon evolutionary thinking altogether. Rather, it is to release that thinking from the poverty of its neo-Darwinian captivity. This requires two steps. One is an enrichment of the concept of the environment within which hominid evolution has taken place. The other is an enhancement of our understanding of the processes that have been at work. When these steps have been taken, we shall be freed from being driven to the construction of implausible Just-so stories, alleging that human capacities of which we have basic experience are totally different in character from what we, in fact, know them to be.

One way of enhancing understanding of the actual scope and character of the human environment can be provided by thinking about the nature of mathematics. Most mathematicians are convinced that their subject is concerned with discovery and not with mere construction.[13] They are not involved in playing amusing intellectual games of their own contrivance, but they are privileged to explore an already-existing realm of mentally accessed reality. In other words, as far as their subject is concerned mathematicians are instinctive platonists. They believe that the object of their study is an everlasting noetic world which contains the rationally beautiful structures that they investigate. Benoit Mandelbrot did not invent his celebrated set; he discovered it. Acknowledgement of the existence of this rational dimension of reality is vital to the possibility of understanding the origin of human mathematical powers.

At some stage of hominid development, our ancestors acquired a brain structure that afforded them access to the mental world of mathematics. It then became as much a part of their environment as were the grasslands over which they roamed. At first, this noetic encounter must have been limited to a utilitarian style of mundane thinking, involving just an engagement with simple arithmetical and geometrical ideas. However, once that intellectual traffic had started it could not be limited to such elementary matters. Our ancestors were beguiled into further exploration of noetic richness which, once begun, continued with an ever-increasing fruitfulness. What drew them into this exploratory process was not a Darwinian drive to the enhanced propagation of their genes, but an entirely different mechanism that we shall consider shortly.

The kind of considerations in the case of mathematical experience that lead us to take seriously an enriched hu-

---

13. See J. C. Polkinghorne, *Belief in God in an Age of Science*, Yale University Press, 1998, ch. 7.

man environment apply equally to other distinctive forms of human ability. Human ethical intuitions indicate the existence of a moral dimension of reality open to our exploration. Our conviction that torturing children is wrong is not some curiously veiled strategy for successful genetic propagation, nor is it merely a convention adopted arbitrarily by our society. It is a *fact* about the nature of reality to which our ancestors gained access at some stage of hominid development. Similarly, human aesthetic experiences gain their authenticity and value from their being encounters with yet another aspect of the multidimensional reality that encompasses humanity. Experiences of beauty are much more than emotion recalled in tranquility; they are engagements with the everlasting truth of being.

Once one accepts the enrichment beyond the merely material of the context within which human life is lived, one is no longer restricted to the notion of Darwinian survival necessity as providing the sole engine driving hominid development. In these noetic realms of rational skill, moral imperative and aesthetic delight—of encounter with the true, the good and the beautiful—other forces are at work to draw out and enhance distinctive human potentialities. Survival is replaced by something that one may call *satisfaction*, the deep contentment of understanding and the joyful delight that draws on enquirers and elicits the growth of their capacities. No doubt the neural ground for the possibility of psychosomatic beings like ourselves to be able to develop aptitudes in this way was afforded by the plasticity of the hominid brain. Much of the vast web of neural networking within our skulls is not genetically predetermined, but it grows epigenetically, in response to learning experiences. It is formed by our actual encounters with reality.

The era of these developments was the time when human culture emerged, generating a language-based Lamarckian ability to transfer information from one generation to the next through a process whose efficiency vastly exceeded the slow and uncertain Darwinian method of differential propagation. It is in these ways that a recognition of the many-layered character of reality, and the variety of modes of response to it, make intelligible the rapid development of the remarkable distinctiveness of human nature.

Hominid evolution inaugurated the exercise of these new creaturely capacities here on planet Earth, but it did not bring into being the reality to which these nascent abilities gave access. What emerged were mathematicians, not mathematics. The latter was always 'there', even if unrecognised by creatures during billions of years of cosmic history. The rational, moral and aesthetic contexts within which hominid capacities began to develop are essential and abiding dimensions of created reality.[14] From the point of view of dual-aspect monism, these realities exist at the extreme mental pole of the comple-

---

14. There may be a contrast here with the ideas of 'emergentist monism'. Philip Clayton says that 'unlike dual-aspect monism, which argues that the mental and the physical are two different ways to characterise one "stuff", emergent monism conceives the relationship between them as temporal and hierarchical' (in T. Peters and G. Bennett [eds], *Bridging Science and Religion*, SCM Press, 2002, p. 116). It is not altogether clear whether this remark is meant to apply simply to the emergence of perceiving entities (in which case there is no conflict with the view taken here), or whether it applies also to what is perceived. The latter notion of emergence would seem to require a constructivist, rather than a realist, conception of the nature of mentally accessed entities. The dual-aspect monism that I espouse seeks to regard the mental and the physical as corresponding to encounters with complementary phases of the 'one stuff' of created reality, rather than simply different characterizations of it.

mentary duality involved, just as sticks and stones exist at the far physical pole.[15]

One should go on to ask what is the origin of these many diverse but interrelated aspects of reality? For the religious believer, the source of these dimensions lies in the unifying will of the Creator, a fundamental insight that makes it intelligible not only that the universe is transparent to our scientific enquiry, but also that it is the arena of moral decision and the carrier of beauty. Those dimensions of reality, the understanding of whose character lies beyond the narrow explanatory horizon of natural science, are not epiphenomenal froth on the surface of fundamentally material world, but they are gifts expressive of the nature of this world's Creator. Thus moral insights are intuitions of God's good and perfect will, and aesthetic delight is a sharing in the Creator's joy in creation, just as the wonderful cosmic order discovered by science is truly a reflection of the Mind of God. Thinking about human experience in this way affords the possibility of a satisfyingly unified account of multilayered reality. Theology can lay just claim to be the true Theory of Everything.

For the theologian, the most important context within which hominid development has taken place is the veiled but grace-giving presence of God. Not only does this ultimate creaturely setting *coram deo*, before the divine, provide the explanation of humanity's continuing encounter with the sacred, but it also gives a distinctive character to theological anthropology. In contrast to many secular accounts of human nature, a Christian understanding of humankind acknowledges

15. J. C. Polkinghorne, *Faith, Science and Understanding*, SPCK/Yale University Press, 2000, pp. 95-9.

our heteronomous dependence on the grace of God, rather than asserting an autonomous human independence. It is the refusal to acknowledge this status of creaturely dependence that is seen as being the root cause of humanity's sinful predicament. Going it alone, 'Doing It My Way', is not the prescription for a truly fulfilled life. Human beings are intrinsically heteronomous. Recently, however, feminist theologians have offered some trenchant criticisms of this concept of human dependency.[16] Clearly there is a danger that heteronomy, wrongly construed, may lead to the picture of a passive and subservient humanity, of the kind that Nietzsche satirised in his famous comment that Christianity is a religion for slaves. No one could deny that this distortion has from time to time been present in the life of the Church, particularly in relation to the subordination of women. Yet, at its best, Christianity has sought not to fall into this error. The long history of theological debate about the correct balance between the roles of grace and freewill is one aspect of the search for a right understanding of heteronomy. Paul gave classic expression to the matter when he exhorted the Philippians (2:12–13) to 'work out your own salvation with fear and trembling; for it is God who is at work in you, enabling you both to will and to work for his good pleasure'. The words of a collect from the Book of Common Prayer, which speaks of God as the one 'whose service is perfect freedom', are another fine expression of a dependence upon the Creator which is the ground of human dignity and worth, rather than the threat of their subversion.

16. For a temperate and helpful account of the issues, see S. Coakley, *Powers and Submissions*, Blackwell, 2002.

CHAPTER FOUR

The Historical Jesus

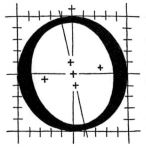NE of the most significant things about Jesus of Nazareth is that we have heard of him. He lived two thousand years ago in a not very important frontier province of the Roman Empire. He died comparatively young, painfully and shamefully executed and deserted by the band of close followers that he had gathered around him. He wrote no book that could have conveyed his message to future generations. Yet everyone has heard of Jesus and, even from a secular perspective, he has been one of the most influential figures in the whole history of the world.

We take our knowledge of Jesus so much for granted that we mostly fail to see how surprising it is. The point can be made by considering comparisons that are sometimes made by scholars with other near-contemporary figures. Jesus spoke with persuasive power to large crowds, but so did John the Baptist. John's followers remained active for some decades after his execution by Herod Antipas (Acts 18:25; 19:2-5), but after that the movement petered out. Today we are only aware of John because of his connection with Jesus. Appolonius of Tyana was a first-century figure credited with miraculous deeds, and the charismatic Rabbis Honi the Circle Drawer and Hanina ben Dosa manifested a striking degree of closeness and familiarity with God.[1] Again there are similarities to aspects of what we are told about Jesus, but none of these people has left an abiding impression on succeeding generations. So why is Jesus different? There is certainly something here that requires an explanation.

If we are to learn about Jesus of Nazareth, it is to the New Testament that we shall have to turn. Near contemporary non-Christian literature only indicates that he lived in Palestine and that his execution took place under the jurisdiction of the Roman Prefect of Judea, Pontius Pilate. That much can be gleaned from the historians Suetonius and Tacitus through their passing references to troubles that arose in Rome associated with the followers of one they call Christus or Chrestus—obviously they did not feel that he was sufficiently important to bother about getting the name right. There is also a brief passage about Jesus in the writings of the Jewish historian and apologist Flavius Josephus, but in the manuscript tradition that has come down to us this has clearly been tampered with by later Christian editors. Any serious attempt to gain knowledge of the life of Jesus has to turn to the four gospels.

The literature relating to the task of investigating the life of Jesus is immense and the conclusions offered are various and

---

1. See G. Vermes, *Jesus the Jew*, SCM Press, 1983, pp. 69-78.

conflicting.[2] The first point to recognise is that the gospels are not biographies in a modern manner, and their writers were not attempting to follow the protocols of modern historiography. The first verse of what is almost certainly the earliest gospel proclaims its subject to be 'The beginning of the good news (*euangelion*) of Jesus Christ, the Son of God' (Mark 1:1). The gospels were written with the intention of presenting Jesus as being the chosen and anointed agent of God's salvific purpose, and the writers' choice of material, and their style of presentation, were directed towards that end. A scientist is familiar with the necessity to interpret data before its significance can become clear. The gospels are interpreted history.

They were probably written during a period of thirty-five to sixty years after the events they describe, based on traditions continuously handed on in a society in which oral communication was a strong and effective manner of conveying and preserving information, quite possibly supplemented by written documents that are now lost to us. Their authors[3] felt free to manipulate the material that had come down to them in ways that assisted them in making their presentation of the significance of Jesus. Apart from the connected accounts of the last week of Jesus' life, the greater part of the material in the gospels consists of actions and sayings of Jesus that appear

2. See, for example, introductory surveys in a textbook style: G. Stanton, *The Gospels and Jesus*, Oxford University Press, 1989; G. Theissen and A. Metz, *The Historical Jesus*, SCM Press, 1998; for a summary of the approach of a careful historian: E. P. Sanders, *The Historical Figure of Jesus*, Penguin, 1993; for a theological approach sensitive to historical questions: N. T. Wright, *The New Testament and the People of God*, SPCK, 1992; *Jesus and the Victory of God*, SPCK, 1996; *The Resurrection of the Son of God*, SPCK, 2003.

3. We do not know for certain the names of the gospel writers, since the customary titles seem to make their first appearance in the second half of the second century, but I shall use the conventional attributions.

to be individual atoms of narration (technically called pericopes) which the gospel writers organised according to the way in which they each wished to present their theme. This freedom of composition means that it is not possible with any reliability to construct a detailed chronological ordering of Jesus' public ministry before that final week in Jerusalem. For example, Mark (2:1-12) puts the story of the paralytic man, let down through the roof to be healed by Jesus, early in his gospel, but Matthew, who probably took the actual story from Mark's account, locates it later (9:1-8), making it part of a collection of healing miracles that he assembles at this point in his narrative.

Writers in the ancient world were not as concerned about accuracy of detail as are modern historians. What mattered to the ancients was to get the main point across. It is instructive to compare three separate accounts given in Acts (9:1-9; 22:6-11; 26:12-18) of the conversion of Paul on the road to Damascus. Did his companions hear the voice from heaven, or was it only Paul? Did they too see the light of the heavenly vision? Did all fall to the ground, or only Paul? Was Paul told straight away what he was called to do, or was he told that this would be made clear to him in Damascus? The different accounts answer these questions in different ways, despite all coming from the pen of the same author. The essence of the story—that Paul saw the risen Christ and was called to his service—is the same in all three versions, and that was all that Luke, the author of Acts, was really worried about.

We should not be surprised or disturbed, therefore, if there are some minor discrepancies between the gospels. Did Jesus heal a blind man on his way into Jericho (Luke 18:35-

43), or on his way out of the city (Mark 10:46-52), or was it two blind men (Matthew 20:29-34)? It is easy to imagine such variations occurring during the period of oral transmission, before gospel writing began. The essential point for the evangelists, when they began their task, was that the deeds of Jesus fulfilled the promise of Isaiah (35:5) that 'the eyes of the blind shall be opened'. However, when we compare the synoptics (that is, Matthew, Mark and Luke) with the fourth gospel (John), a much greater degree of difference becomes evident.

There is an immediately apparent contrast in tone and style. In the synoptics Jesus speaks in crisp parables, which often centre on the Kingdom of God and whose images are earthed in the life of first-century Palestine. In John, however, there are long speeches, centring on Jesus himself—'I am the bread of life' (6:35), 'the light of the world' (8:12), 'the good shepherd' (10:11) and so on. The Johannine discourse has a hauntingly timeless quality to it. It is difficult to think that during his lifetime Jesus could actually have spoken in two quite such different ways. Many believe that John wrote his gospel at the end of a long life of deep reflection on the inner significance of Jesus, expressing his insights in discourses that he then attributed to Jesus. This would be a perfectly acceptable practice in the ancient world, in accord with its conventions for conveying the most profound truth. Plato puts many of his thoughts into the mouth of Socrates, in whose tradition he stood, and something not altogether unlike this is still going on in Shakespeare's historical plays. Commenting on Jesus' statement 'I am the true vine' (John 15:1), E. P. Sanders says that John 'would not have agreed that historical accuracy and truth are synonymous, any more than he thought the true vine was a vegetable'.[4] However, there are also occasional hints in the synoptics of themes that in a more highly developed form are such a striking a feature in John. For example, Jesus sometimes speaks of himself as one who has been sent (as in Mark 9:37), and there is the so-called Johannine thunderbolt in Matthew (11:27) and Luke (10:22), where Jesus says, 'No one knows the Son except the Father, and no one knows the Father except the Son' (though some have argued that this is really a kind of proverb, interpreting it in the generic sense that only a son knows a father, or a father a son). It is possible to exaggerate the differences between John and the others.

Nevertheless, it is certainly a more tricky matter to attempt to use John as an historical source for the life of Jesus than it is in the case of the synoptics. Yet, having acknowledged that, it would also be a mistake to assume that there is no historical information to be culled from the fourth gospel. Of the four it shows the most concern for marking chronological and geographical locations. Yet there are also perplexing clashes with the synoptic tradition. John places the incident of Jesus' cleansing of the Temple at the beginning of his public ministry, while the synoptics locate it at the start of the final week in Jerusalem. There is a celebrated disagreement between John and the rest about the date of the crucifixion. All agree that it was on a Friday, but John dates it on the fourteenth of the Jewish month of Nisan (the day of Preparation of the Passover, when the lambs were slain in the Temple), while Matthew, Mark and Luke date it 15 Nisan (the day that began with the evening Passover meal). Many scholars think

---

4. Sanders, *Historical Figure*, p. 71.

it is likely that John's dating has been influenced by symbolic considerations: Jesus, the Lamb of God (John 1:29), dies at the same hour that the Passover lambs are being slaughtered.

At other times, there are reasons for preferring the detail of the Johannine account. An interesting case is presented by the various stories of the miraculous feedings of the multitude that are given in all four gospels (Mark 6:35-46; 8:1-10; with parallels to one or both in Matthew, Luke and John). Of all the miracle stories attributed to Jesus, these are the most difficult for which to envisage what it might have been that happened. In consequence, these incidents are very seldom represented in Christian art. Sanders draws our attention to an aspect of the accounts that it would have been easy to miss.[5] Jesus' healing miracles are often said to be greeted with astonishment by the spectators, and so it is that his fame came to be spread abroad (Mark 1:28, etc). Yet the stories of the feedings end abruptly, with the crowd being dismissed without their having had a chance to utter any comments and the disciples being smartly despatched in advance of Jesus himself. It seems as if the evangelists are trying to suggest that these extraordinary happenings occurred without occasioning an outburst of excitement and notoriety. Only when one reads John's account does one gain a clue to understanding this apparent oddity. He alone tells us that 'Jesus realised that they were about to come and take him by force to make him king' (John 6:15), and that is why he hastily closed things down. Jesus was not a this-worldly revolutionary of the Zealot type, and when the evangelists came to write their accounts they were still anxious to make sure this misconception did not get around. I believe

5. Ibid., p. 156.

that is why Matthew, Mark and Luke did not write any crowd reactions into their versions.

But did the feedings actually happen? Trying to answer that question is complicated by the fact that the stories carry a high symbolic significance (though this is not explicitly alluded to by any of the gospel writers). One of the expected signs of the fulfilment of God's final purposes was the Messianic Banquet (Isaiah 25:6), and the feeding stories would carry a strong overtone of being anticipations of that consummatory event. It is also very probable that these stories were treasured and repeated within the early Christian community partly because of their association with the regular experience of the Eucharistic meal in the Church's worship (though bread and fish are not a precise parallel to bread and wine). Attempting to assess the historicity of the feedings immediately brings into focus the need to attend to the issue of the kind of explanatory principles that may properly be allowed to control the discussion. A writer making a purely secular historical judgement, based on the premise of natural uniformity, so that what usually happens is what always happens, will find the feeding stories incredible as actual occurrences. Either they must be powerful symbolic tales that got into the tradition as if they were actual happenings, or they must be explained away by rather feeble rationalisations, such as the idea that Jesus inspired people to produce and share the picnics that they had been keeping hidden for their private consumption. Yet the whole of the New Testament is predicated upon the claim that in Jesus something extraordinary and without precedent was actually happening. In the view of those early Christians, God was present in him, and in his words and deeds, in some entirely new and permanently significant way. This claim is asserted to have an anchorage in history, but to involve events whose proper interpretation demands taking into account the unique divine dimension of what was going on. According to this understanding, in the life of Jesus we have to do with happenings that cannot adequately be understood in terms of the categories of secular history alone. A *theological* perspective is claimed to be absolutely essential to do justice to what is involved. If this point of view is accepted, it makes it a conceivable possibility that there could have been unprecedented occurrences.

In that case, an inescapable circularity enters into the argument. If there was something truly unique about Jesus — going beyond recognisably familiar categories such as itinerant wonder-worker, prophetic figure, or charismatic teacher — then at least it becomes conceivable that his life manifested powers and happenings of wholly novel kinds. If Jesus' life had these extraordinary features, it becomes likely that there was something altogether exceptional about him. Is the resulting circularity vicious or benign? Scientists know that in their disciplines theory and experiment are inextricably entangled (theory is needed to interpret what it is that experiments are actually measuring; observations confirm or disconfirm theories). The long-term fruitfulness displayed by scientific understanding is persuasive that in this case the circle is benign. How is it the case with Jesus research?

An answer can only be attempted in the light of a careful and scrupulous investigation of the evidence. For more than two centuries the New Testament has been subject to intense scholarly scrutiny. Many different interpretations of the material have been proposed and 'assured results' that are very widely accepted are in somewhat short supply. Obviously a brief chapter like this cannot begin to be adequate to survey the complexity and importance of the discussion. What I am principally trying to do here is to sketch what I believe a truth-seeking reader, neither inclined to too-great credulity nor to too-great scepticism, might conclude on the basis of an initial investigation that accepts a self-denying ordinance of not bringing theological considerations into play at the first stage of the discussion. I shall then suggest that following out this strategy raises, but of itself cannot answer, further questions that it would seem necessary to address if one is to form an adequate understanding of Jesus. It is my belief that completing the task then requires allowing theological insights and evaluations to come into play, with the attendant possibility that unprecedented events may need to be considered as being a part of what actually happened.

Let us return to the synoptic gospels. First one must acknowledge that it is very likely that not all the words that they attribute to Jesus were actually spoken by him. An instructive example involves the Jewish food laws. In Mark 7:14-23, there is a passage in which Jesus declares to the crowd that 'there is nothing outside a person that by going in can defile, but the things that come out are what defile'. As is often the case in Mark, the disciples find this saying enigmatic and so they ask Jesus privately what he meant by it. He says, 'Do you not see that whatever goes into a person from outside cannot defile, since it enters not the heart but the stomach, and goes out into the sewer'. Just in case the reader still does not get the point, Mark adds an editorial comment, 'Thus he declared all foods clean'. Yet Acts and the Pauline epistles make it abundantly clear that for many years in a period antedating the appearance of Mark's gospel, there were contentious arguments in the Christian community about whether Jewish food laws had any relevance to its way of life. If Jesus had spoken so plainly in his lifetime, it is difficult to see how any uncertainty could have remained. Also it is highly likely that, if he had actually dissented so uncompromisingly from such a central aspect of Jewish life, he would there and then have become embroiled in a severe confrontation on the issue with the authorities. It seems very likely indeed that Mark, or someone else before him, prayerfully reflecting on the troublesome issues of a later day, came to the conclusion that this is what Jesus would have said had the matter arisen then, and so the words became incorporated into the gospel account (possibly linked with the preserved recollection of a puzzling remark that Jesus had actually made). So much seems likely to me, but it also is clear that the gospel writers did not exercise this creative freedom in an unrestrained way. An even more contentious issue for the early Church was whether male Gentile believers in Jesus needed to receive the Jewish rite of circumcision. Yet no word is found in the gospels conveniently settling the matter.

I believe that the gospel writers were seeking to tell things as they were, according to the knowledge that had come to them, and within the accepted narrative conventions of their day. The honesty of their intent is signified to me by the way in which they record incidents and words of Jesus that must have given them great difficulty, but which had to be on record as part of what had actually happened. One incident that gave rise to perplexity, but which is reported in all four gospels, is the baptism of Jesus by John. After all, John's baptism was a washing of repentance, and the early Church did not think that Jesus had sins to confess and renounce, so there was some anxiety about what it all meant. Another incident that all the gospels record is the discreditable story of Peter's denial of Jesus during the frightening events of that Thursday night of the arrest. Only a strong desire to tell the truth as it had actually happened could have led to the propagation of so disreputable a tale about an outstanding leader of the early Christian community. Matthew (8:22) and Luke (9:60) both quote a remark of Jesus that would have shocked both Jew and Gentile alike in the ancient world, where proper burial was a sacred duty, 'Let the dead bury their own dead'. When Matthew came to compile his gospel, it must have been as clear to him then as it is to us today, that the word of Jesus to his disciples, 'you will not have gone through all the towns of Israel before the Son of Man comes' (10:23), had not received a straightforward kind of fulfilment, yet it had to be reported. And then there is the cry of dereliction from the cross, 'My God, my God, why have you forsaken me?'. So strange and terrible is this cry that Mark (15:33) gives also the actual Aramaic words that Jesus would have uttered. Matthew (27:46) tones it down a little by turning the Aramaic into Hebrew, thereby suggesting a quotation from Psalm 22. It seems to me that both evangelists have to report this disturbing cry because it is part of the actual story of Jesus' passion. The gospels are not hagiographies, seeking to present as pleasing and unproblematic a picture as possible. Their concern is to tell what is most important and significant in the words and deeds of Jesus in as essentially truthful a fashion as the evangelists can manage.

Part of the task of assessing the life of Jesus clearly requires reaching decisions about which words that the gospels attribute to him were actually uttered by him, rather than being subsequent creations of the first generation of Christians. Needless to say, a great variety of scholarly answers have been proffered, ranging from 'almost nothing' to 'practically everything'. Clearly, absolute certainty is not going to be attainable in the case of Jesus, any more than it could be for any other figure in history, but one may seek to reach conclusions with as high a degree of probability as possible. One technique employed has been to try to devise quasi-scientific 'objective' criteria for making decisions.

One criterion of this kind that has been used is 'double dissimilarity'. If a saying has no obvious connection with the Judaism of his day, or with the life and concerns of the early Church, then it must surely have come from Jesus himself. It is not entirely straightforward to apply this test, since our knowledge of first-century Judaism, and of the early Christian community, is fairly fragmentary, but one can see that if the criterion appears to be satisfied, that does constitute a persuasive indication of authenticity. However, one must expect that only a small fraction of Jesus' actual sayings would pass this test. He surely cannot have been free from the impress of his Jewish heritage, nor could he have been without a lasting influence on his followers. Applying the test of double dissimilarity as the sole criterion of the authentic thought of Isaac Newton would lead to dismissing his work on mechanics and gravitation altogether, because of its connections to the preceding work of Galileo and Johannes Kepler and its continued shaping of theoretical physics down to the present day.

Another criterion of a quasi-scientific type has been 'multiple attestation'. If a saying can be seen to be derived from two or more independent sources, that naturally strengthens a claim to authenticity. Once again, one can acknowledge the positive value of passing this test without believing that it is indispensable. The amount of source material available for the life of Jesus is limited, even if it is much greater than is the case for many well-known figures in the ancient world. In consequence, where multiple attestation occurs it is something of a fortuitous luxury. It also seems to me that those scholars who have been most keen to use this test as a prime resource have not always been very consistent in their application of it. At times they can seem quite rash in their appeal to dubious extra-canonical sources,[6] while at other times they display an undue degree of suspicion of the New Testament material. It is not uncommon for such scholars to question the story of the empty tomb, despite its being given in all four gospels (with the sort of minor discrepancies about the number and names of the women, and the precise hour of early morning at which they paid their visit, of the kind that we have seen are not really significant in assessing reliability). Of course one has to be careful to count only sources that are reasonably likely to be independent. It is quite probable that Matthew and Luke drew their accounts of the empty tomb at least partly from that of Mark, but the tradition behind John can plausibly be considered independent of the synoptics, so that there is at least double attestation in this case.

Personally I am sceptical that these scientific-sounding criteria can be more than moderately useful. They leave the greater part of the gospels' material unassessed, but this evidence should certainly not be written off because it cannot be made to jump through these particular hoops. I believe

6. See, for example, J. D. Crossan, *The Historical Jesus*, HarperCollins, 1992. In a quest for sources, Crossan is prepared to place what seems to be a quite remarkable reliance on non-canonical writings, such as the gnostic *Gospel of Thomas* and the apocryphal *Gospel of Peter.*

that the careful investigator should be willing also to use more qualitative procedures. A critical question to address is whether the gospel material taken as a whole carries the impress of a single remarkable figure lying behind most of it. My answer would be, Yes. Matters such as the powerful impact of the parables, and a distinctive ability in controversy to point the conversation straight to the heart of the matter by asking a further question in response to a question (see, for example, Mark 12:13–27), seem to me to indicate a single original mind at work, rather than their being the result of a variety of anonymous early Christians making up stories. I do not think that all is totally clear, or that Jesus' character and convictions are absolutely transparent to us. Yet, for me there is something mysterious yet hopeful, at times perplexing, at times enthralling, frequently inescapably challenging, about the figure of Jesus that emerges from the gospel pages. He is someone whom I absolutely have to take into account with complete seriousness in my exploration of reality. I want to concur with the verdict of a scrupulous and distinguished New Testament scholar, Charlie Moule, when he wrote, 'the general effect of these several more or less impressionistic portraits [in the gospels] is to convey a total conception of a personality, striking, original, baffling yet ultimately illuminating'.[7]

The most characteristic action attributed to Jesus in the course of his public ministry is the healing of the sick. It would be impossible to excise these many stories from the gospel record without unacceptably mutilating the documents.

7. C. F. D. Moule, *The Origin of Christology*, Cambridge University Press, 1977, p. 156.

Sanders, in his cautious manner as an historian, says, 'I think we can be fairly certain that initially Jesus' *fame* came as a result of healing, especially exorcism'.[8] That final comment reminds us that in the synoptic gospels the manner of these healings is often represented as being due to the expulsion of 'demons'. A first-century Palestinian like Jesus would naturally think and operate within the idiom of his day. Though there are also healing stories in John, it is interesting that the fourth gospel never uses the language of exorcism. In the idiom of our own day, we can understand some of the healing encounters with Jesus in terms of the psychosomatic influence of an exceptionally gifted charismatic healer, though this would scarcely apply to stories of healing at a distance, such as Matthew 8:5–13, with parallels in Luke and John.

It used to be popular to suppose that Jesus preached a message of love and forgiveness, totally opposed to the narrow and rigid legalism of his contemporaries in Judaism, especially the Pharisees, and that the hostility towards him that eventually led to his death arose from this confrontation. Recent research into the first-century Jewish background of the life of Jesus has persuaded a number of contemporary scholars that this view is untenable.[9] Many Jews of Jesus' day seem to have embraced an understanding of God's relationship to themselves that corresponded to what may be called 'covenantal nomism', the concept that within the community of the *Torah* (the Law), gracious acceptance was freely available to the repentant sinner. This idea would have been familiar to the Pharisees, and within its broad category there would have

8. Sanders, *Historical Figure*, p. 154.
9. See especially E. P. Sanders, *Jesus and Judaism*, SCM Press, 1985.

been scope for a variety of different detailed understandings of its implications. The picture of first-century Judaism now given us by scholars is one of a religion that tolerated reasonable diversity of interpretation, so that it would have been unthinkable to seek someone's death simply on the grounds of a disagreement about these matters.

Two comments may be made. One is that one would expect, even if this general understanding is correct, that there would have been considerable variations of attitude on the part of individuals. I find it difficult to believe that the stories of bitter confrontations between Jesus and at least some of the Pharisees that one finds in the gospels are *all* retrojected images of the disputes that certainly did arise between Judaism and nascent Christianity towards the end of the first century. The second and more important point is to remember that scholars continue to agree that within mainstream Jewish thinking, acceptance of sinners back into a covenantal relationship with God required prior repentance for their previous way of life. There is good evidence that Jesus scandalised his pious contemporaries by his willingness to accept highly dubious people, such as tax collectors and prostitutes, and even to eat with them, without exacting the prior condition of publicly expressed penitence (for example, Mark 2:13-17 and parallels).

When he acted in this way, it appears that Jesus relied on exercising his own understanding and authority as guides to what was in accordance with God's will. Another illustration of Jesus' remarkable self-authenticating confidence in his own decisions on religious matters is to be found in the Sermon on the Mount. The verses in which he comments on the *Torah* (Matthew 5:17-43) do not set the Law aside—in fact

Jesus explicitly says, 'Do not think that I have come to abolish the law or the prophets; I have come not to abolish but to fulfil' (v. 17). In the passages that follow he takes various commandments (against murder, adultery, and so on) and deepens their application to take into account internal attitudes (no anger, no lust, and so on), without in the slightest abrogating their force. Although these sets of verses are often called 'the Antitheses', they do not set up an opposition between Jesus and the Jewish Law. As far as their content was concerned, they would have been unlikely to have roused the hostility of his contemporaries, but what would have been likely to provoke opposition was the manner in which Jesus justified his conclusions. Here was not a case of respectful Rabbinic argument about details of interpretation, but a categorical assertion of the speaker's right to settle the matter. The repeated refrain is 'You have heard it was said to those of ancient times . . . But I say to you . . .' (vv 21–22; 27–28; 31–32; 33–34; 38–39; 43–44). When one recalls that the words of ancient times were given by God to Moses on Mount Sinai, one sees that a truly remarkable claim to authority is being made by Jesus, and it is possible to understand how many pious Jews might have found this very offensive. Sanders comments that Jesus 'regarded himself as having full authority to speak and act on behalf of God'.[10] When John has Jesus say 'Whoever has seen me has seen the Father' (14:9), this seems to be another case of the explicit intensification of something already implicit in the synoptics. All four gospels portray Jesus as frequently prefixing his assertions with the Semitic word 'amen', essentially affirming the unshakeable certainty of what is to follow. (In

10. Sanders, *Historical Figure*, p. 238.

English versions the force of this idiosyncratic usage is unfortunately often disguised from the reader by the use of some such word as 'truly' in the translation).

One of the points on which the great majority of New Testament scholars can actually agree is that a central concern of Jesus' teaching was the Kingdom of God (that is, the rule of the divine will and the fulfilment of the divine purpose in human lives). In the synoptics, many of the parables are concerned with the nature of that kingdom and its coming. In John, the phrase occurs only twice, in chapter 3 (and there is also reference to Jesus' kingdom in chapter 18), but the often-repeated phrase 'eternal life' plays a somewhat similar role in the idiom of that gospel. In the synoptics, Jesus' acts are seen as signs of the Kingdom (for example, Luke 11:20). Sometimes the Kingdom is spoken of in terms of its being present and sometimes its coming is said to be something to be expected in the future. These two possibilities can occur side by side in the gospel narrative (compare Luke 17:21 with Luke 17:22–25). Future expectation is often linked with the coming of an eschatological figure called the Son of Man. The correct interpretation of this phrase has probably given rise to more scholarly dispute than any other topic in New Testament studies.

The words are as odd in Greek as they are in English, but in a Semitic language such as the Aramaic of Jesus' day, they can simply be a perfectly natural way of saying 'human' (just as 'children of Israel' was an accepted Semitic way of saying 'Israelites'). However, the phrase is also used in the Book of Daniel (7:11–18) to denote a specific heavenly figure who appears before God in connection with the vindication of the persecuted 'saints of the Most High'. To add further complication to the question of possible meanings, 'son of man' seems also to have been usable in the first century as an impersonal circumlocution for the speaker, rather as 'one' can be used in English.

In the New Testament, where the phrase is always given in Greek and never in transliterated Aramaic, its use is almost entirely confined to the four gospels and in them there is only one trivial occasion (John 12:34) when it is placed on the lips of anyone other than Jesus himself. There is no evidence to suggest that the phrase was in common usage in the early Christian community as a title it had conferred on Jesus. Rather, it seems that it must be to him that we have to look if we are to determine its significance.

In the gospels, there are certainly examples of its impersonal use (for example, Matthew 16:13, where in the parallel verses Mark 8:7 and Luke 9:18, it is replaced by 'I'), and perhaps also as a synonym for human being (possibly so in Mark 2:28). Yet it is also used in a way that makes it appear to be attached with unique significance to Jesus himself (for example, Mark 8:31), and at other times as relating to a figure closely connected with Jesus but not unequivocally identified with him (for example, Mark 8:38). The possible ambiguity of these latter sayings assures us that they go back to Jesus himself, since the post-resurrection church would not have been in any doubt about the matter. I personally believe that Jesus used the title Son of Man for himself as a way of characterising his role in God's plans and that this usage was in the light of Daniel 7.[11] This would certainly imply that Jesus had a very high view of his own unique role in the final fulfilment of the divine purposes.

11. J. C. Polkinghorne, *Science and Christian Belief/The Faith of a Physicist*, SPCK/Fortress, 1994/1996, pp. 98-100.

Consideration of eschatological expectations, and what Jesus believed would be his part in them, leads us to issues that can begin to seem relevant to attempting to answer the question of why there was opposition to Jesus that became so intense that it led to his being condemned to death. A claim to play a unique role in Israel's future and in the final fulfilment of God's plans would have been an unacceptable self-aggrandisement in the eyes of many of his Jewish contemporaries. Argument about detailed interpretation of the Law was one thing; putting oneself centre stage in God's sight and acting as a kind of viceroy for the divine Kingdom (so that Jesus could accept sinners simply on his own authority) was quite another. Sanders comments that 'Although he did not oppose the law, he did indicate that what was most important was accepting him and following him'.[12] In particular, eschatological claims made by Jesus were particularly offensive. Tom Wright says that 'the clash between Jesus and his Jewish contemporaries, especially the Pharisees, must be seen in terms of *alternative political agendas* generated by *alternative eschatological beliefs*'.[13]

Scholars today believe that a particularly critical and provocative incident, giving dramatic public expression to these different political and eschatological understandings, was Jesus' act of cleansing the Temple after his entry into Jerusalem at the start of the last week of his life (Mark 11:15-19 and parallels, accepting the chronology of the synoptics against that of John). This was something much more than just a protest against commercialism, for it carried the significance

12. Sanders, *Historical Figure*, p. 236.
13. Wright, *Victory of God*, p. 390.

of 'an acted parable of judgement'.[14] The last straw for the High Priest Caiaphas and his associates was that Jesus had now chosen to extend his self-authenticated exercise of authority to the sacred precincts of the Temple itself. In their eyes he would have seemed to be guilty of troubling the people by such idiosyncratic judgements on the established way of life of his fellow Jews. Worse would follow if this reckless action roused the volatile crowds, always present in Jerusalem at Passover time, to civil disturbance and riot. After a brief informal meeting,[15] the chief priests handed Jesus over to the Roman Prefect, in the expectation that Pilate would take the easy way out for preserving order and have him executed. Practical men knew that it was better 'to have one man die for the people than to have the whole nation destroyed' (John 11:50). In the end Pilate agreed, and he handed Jesus over to be crucified. It seems that he did not really believe that Jesus was a potentially dangerous insurgent, for if he had he would surely have hunted down the disciples in their places of hiding and eliminated them too. Pilate simply acted in the way he judged would best preserve the stability of the tiresome province he had been given to govern.

Jesus died a painful death. He died a shameful death, one reserved for slaves and rebels, and one abhorred by any pious Jew as a sign of God's rejection, since Deuteronomy (21:23) proclaimed that 'anyone hung on a tree is under God's curse'. Jesus died deserted by his followers (except for a small band of brave women), and with a cry of dereliction on his lips. On

14. Ibid., p. 416.

15. Sanders is cautiously inclined to prefer John's account of an irregular private cabal to the synoptics' story of a formal hearing before the Jewish Council; see *Historical Figure*, p. 72.

the face of it, his life ended in utter failure.[16] The only reasonable expectation would be that he then disappeared from history into the black hole reserved for those who fail because, in the end, they could not sustain the grandiose pretensions about themselves that they had assumed during their lifetimes. (Palestinian history in the decades following Jesus contains examples of such deluded people, whose names are now familiar only to historians of the first century.) The claim to a special authority to speak and act on God's behalf must have seemed pretty pitiable to anyone looking at that rejected man, dying on a gibbet outside Jerusalem.

Yet we have all heard of Jesus. We should not rest content with any account of him that does not make this fact intelligible. The story of Jesus continued after his death. The followers who had deserted him in terror and disappointment, very soon were standing up in Jerusalem, proclaiming that God had made him 'both Lord [a title particularly associated with the one true God of Israel] and Messiah' (Acts 2:36). They took that faith right across the known world and many of them lost their lives rather than deny it when they were persecuted. Their confident proclamation has come down through the centuries to us today, set out in the writings of the New Testament, and it is supported by the continuing witness of the Church. From the first, the origin of that continuing story has been asserted to lie in the fact that God raised Jesus from the dead the third day after his death and burial.

Clearly *something* must have happened of very great mag-

16. This is in striking contrast to other great founder figures of religious traditions. Moses, the Buddha and Mahommed all end their lives in honoured old age, surrounded by their disciples who are resolved to carry on the work and message of the Master.

nitude to produce this astonishing sequel to the earthly life of Jesus. I believe that it must have been very much more than just a dawning recognition that the message of Jesus could carry on even if he were dead, coupled with a resolve to seek to preserve his memory. A purely secular approach to the life of Jesus seems to me to prove inadequate at this point. With characteristic caution and candour, Sanders says, 'That Jesus' followers (and later Paul) had resurrection experiences is, in my judgement, a fact. What the reality was that gave rise to those experiences I do not know'.[17] It is difficult for an historian working within the limits of secular protocols to say more by way of a verdict. I think that to understand what happened one must be prepared to go beyond the confines of such secular thinking, for I believe that the source of the continuing story of Jesus was exactly what those first Christian followers claimed it to be, 'This Jesus God raised up, and of all this we are witnesses' (Acts 2:32).

The resurrection is the pivot on which Christian belief turns. Without it, it seems to me that the story of Jesus' life and its continuing aftermath is not fully intelligible. There are two kinds of evidential claims recorded in the New Testament that need to be considered in pursuit of a case for belief in the resurrection. One relates to the appearances of the risen Jesus. The earliest testimony is that of Paul in his first letter to the Corinthians (15:1-8). Writing about the year 55, he reminds them of what he had passed on to them when he had founded the church in Corinth a few years earlier,

> For I handed on to you as of first importance what I in turn had received: that Christ died for our sins in accor-

17. Sanders, *Historical Figure*, p. 280.

dance with the scriptures, and that he was buried, and that he was raised the third day in accordance with the scriptures, and that he appeared to Cephas, then to the twelve . . . (vv 3–5)

When Paul speaks about what he himself had received, this seems very likely to be a reference to what he had been taught following his conversion on the road to Damascus. This would take the testimony back to within two or three years of the crucifixion, and some confirmation of this antiquity can be found in the actual wording of the text (for example the use of the Aramaic 'Cephas', rather than Peter). The whole passage is very condensed, ending with a list of those said on various occasions to have seen the risen Christ, the last of whom is Paul himself.

If one wants to gain some idea of what these appearance experiences might have been like, it is necessary to turn to the gospels. A very varied picture then confronts us, in quite striking contrast to the degree of similarity in the stories that the evangelists tell us about the last week of Jesus' life. In Mark, though there are two predictions that the risen Jesus will meet with his disciples in Galilee (14:28; 16:7), the authentic text that has come down to us ends at 16:8 with the women's fear at the discovery of the empty tomb, and there is no explicit account of an appearance. (The stories given in some English versions of the end of Mark derive from second-century manuscript additions that do not constitute a reliable primary source.) In Matthew, Jesus appears to the women after the discovery of the empty tomb, but the main appearance event is located on a hillside in Galilee. In Luke, everything seems to happen in Jerusalem on the first Easter Day, though the same author when writing Acts (1:3) speaks of a sequence of appearances over a forty-day period. In John, there are appearances both in Jerusalem and in Galilee.

At first sight, such variety might make one think that one was dealing with a gaggle of independently made-up tales. However, there is a surprising common feature in the stories that persuades me that this is not the case. This feature is the recurrent theme that initially it was difficult to recognise the risen Christ (Mary Magdalene mistakes him for the gardener; the couple on the road to Emmaus only recognise who has been with them at the very end of their encounter; Matthew quite frankly tells us that some on that Galilean hillside 'doubted' who it was [28:12]; and so on). One may speculate exactly why this was so, but it seems very likely indeed to me that this emphasis on a moment of disclosure not reached without difficulty is a genuine historical reminiscence of what these meetings were like, and very unlikely that it would be a fortuitous common factor in a bunch of stories concocted independently by several different authors.

Then there are the stories of the empty tomb given, as we have already noted, in essentially the same form in all the four gospels. It is important to remember that in the New Testament the empty tomb is not presented as a piece of knockdown evidence. On the contrary, the first reaction of the women is said to be fear and perplexity (Matthew 28:5; Mark 16:5–8; Luke 25:4–5; John 20:2), and angelic interpretation is necessary before the true significance of what has happened dawns on them.

There are two difficulties about the empty tomb stories, however, that have to be addressed. The first is why the emptiness of the tomb is not mentioned by Paul, if it really happened and it was so important. However, what we have from Paul is a collection of occasional letters, rather than a considered exposition of all that he knew, and this makes an argument from silence unpersuasive in his case. Many of us also think it is significant that Paul, in that very spare account in 1 Corinthians 15, took space to note that Jesus 'was buried'. A first-century Jew like Paul, taking a psychosomatic view of human nature, would have been very unlikely indeed to have believed that Jesus lives (as he undoubtedly did believe), while his body still mouldered in a tomb.

But was there actually an identifiable tomb? Certainly it was a frequent Roman custom to bury executed felons (which is what Jesus would have seemed to the authorities to be) in a common grave. However, there is archaeological evidence from the first century which shows that this practice was not without exception. The story of a separate tomb for Jesus gains credibility from its association with Joseph of Arimathea. There seems to be no reason why the early Church should have assigned to this otherwise unknown figure the very honourable role of burying Jesus, unless he actually performed it. In the long argument between Judaism and the Church about the significance of Jesus (a history of contention that can be traced back to the first century), it is always common ground that there was a tomb and that it was empty. What was disputed was the reason for this: the resurrection or an act of deliberate deceit on the part of the disciples? The latter is surely incredible. Men do not die for what they know to be a lie.

Perhaps the strongest reason for taking the stories of the empty tomb absolutely seriously lies in the fact that it is women who play the leading role. It would have been very unlikely for anyone in the ancient world who was concocting a story to assign the principal part to women since, in those times, they were not considered capable of being reliable witnesses in a court of law. It is surely much more probable that they appear in the gospel accounts precisely because they actually fulfilled the role that the stories assign to them, and in so doing they made their startling discovery.

Two further circumstantial points can be made. One is the fact that early on the Christian community came to recognise Sunday, rather than the Jewish Saturday, as its special day of worship. The testimony is that this was because it was known to be the day on which the Lord had risen from the dead. The other point is to note that the characteristic way in which the Christian Church has always spoken about Jesus is as a living Lord in the present, rather than a revered Founder Figure in the past.

We have surveyed the evidence that can be interpreted as anchoring the claim of the resurrection of Jesus in historical testimony, even if the event itself transcends the simply historical. The grounds for belief that these considerations represent, taken together with the need to understand how the influence of Jesus continued against all the apparent odds, are sufficient to persuade me of the truth of Jesus' resurrection. Yet I recognise that precisely how one weighs the evidence will depend upon one's world view of the nature of reality, and in particular on whether one finds it possible to believe in the existence of a God who acts in history. No one can be forced to go beyond the kind of scepticism so trenchantly expressed by David Hume about any whiff of the miraculous, but neither can anyone be denied the rational possibility of a theological interpretation that affirms God's unique action in Jesus Christ. As with all metaphysical issues, the final decision has to be reached on metaphysical grounds, such as coherence and scope of understanding, together with an assessment of the significance of the unique in addition to the role of what is general. The theological grounds for my own belief in the resurrection have been given elsewhere.[18] This chapter concentrates on a search for historically motivated considerations that can raise, but not of themselves finally settle, these deeper issues and more profound questions.

Suffice it to say, that if a person believes in the resurrection—or at least is open to the possibility of that belief—then a simply secular reading of the gospels will no longer be sufficient. Their adequate assessment cannot be grounded solely on the assumptions that what usually happens is what always happens, and that Jesus can find satisfactory classification within the already-known categories of types of religious figures. Something absolutely significant and absolutely unique must have been going on in his life if he alone was the person whom God raised from the dead within history, to enter into an unending and glorified life within the divine presence. A theological approach is then an indispensable part of seeking to understand what that unique significance of Jesus actually is. This process of theological reflection on the life of Jesus begins in the pages of the New Testament itself, starting with writings such as the Pauline epistles that antedate the gospels. The New Testament writers, despite being monotheistic Jews, found themselves driven by their experience of the risen Christ to use theistic-sounding language about him, as when Paul bracketed God and Jesus together in his characteristic greeting 'Grace to you and peace from God our Father

18. Polkinghorne, *Christian Belief/Faith*, ch. 6.

and the Lord Jesus Christ' (Paul applies the title 'Lord' to Jesus more than two hundred times in his letters), or when he quoted an Old Testament passage about the God of Israel and, without apparently batting an eyelid, could apply it to Jesus (for example, compare Philippians 2:10–11 with Isaiah 45:23). Mature reflection on these phenomena eventually led the Church to a high doctrine of Christology. I have written in some detail on this subject elsewhere.[19] Let me now just remark that many people ask, If there is a God, why have God's existence and the divine nature not been made known more clearly? The Christian reply is that, though something can be known of deity from the kind of general considerations that traditionally are the concern of natural theology,[20] the trinitarian God can only properly be revealed through the uniqueness of particular historical events. True meeting with the divine has to have the character of personal encounter rather than of accessing a celestial Web site. At the heart of Christian belief is the mysterious and exciting idea that the invisible God has in fact acted to make the divine nature known in the most humanly accessible way, in the life, death and resurrection of Jesus Christ.

It was its high Christology that led the Church to develop its trinitarian understanding of the nature of God, and that is a subject that I want to address in the next chapter.

---

19. Ibid., ch 7; *Belief in God in an Age of Science*, Yale University Press, 1998, ch. 2.

20. See, for example, Polkinghorne, *Belief in God*, ch. 1.

CHAPTER FIVE

Divine Reality: The Trinity

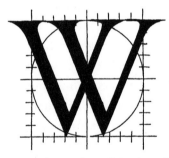RITING soon after the great initiating discoveries of quantum theory, the biologist J. B. S Haldane said that he suspected that the universe is not only queerer than we suppose, but queerer than we can suppose. A similar remark, to the effect that the world has proved stranger than we could have thought, has been a kind of free-floating *logion* in the folk tradition of quantum physicists, attributed, amongst others, to Bohr and to Heisenberg. Certainly, quantum theory strikingly illustrates the limits of commonsense thinking. Even our notions of logic have had to be reassessed in the light of quantum reality. Classical logic, of the kind familiar to Aristotle and to the legendary man on top of the Clapham omnibus, is based on the stark choice enforced upon us by the law of the excluded middle: there is no term intermediate between 'A' and 'not A'. But we have seen that in the quantum world there is a whole range of states in which an

electron is in a superposed mixture of 'here' (A) and 'there' (not A). The existence of these hitherto undreamed-of middle terms implies that a new kind of 'quantum logic' is required for entities such as electrons.

There are limits to the degree of understanding attainable at any given time in physics. Coherence is surely an ultimate essential in a fundamental theory, but sometimes it may be necessary for a while to be content to leave problems of consistency not fully resolved. The two greatest twentieth-century discoveries in physics were quantum theory and general relativity. We have seen that for almost eighty years physicists have had to live with these two theories imperfectly reconciled with each other, since the straightforward attempt to put them together leads to nonsensical infinities. Present hopes for remedying this state of affairs centre on the ingenious, but highly conjectural, theory of superstrings,[1] apparently lying far beyond any possibility of direct experimental verification and requiring for its consistency that one adopts the hypothesis that there exist six or seven inaccessible dimensions of spacetime in addition to the four of our immediate experience. String theory asserts that what experimentalists seem to encounter as elementary particles are, in fact, the effects of tiny loops, 10^{-33}cm across, vibrating in that multidimensional space. (Who said that scientists do not believe in unseen realities?)

Full understanding of the implications of a fundamental scientific theory is clearly something much to be desired. Yet we have seen that the measurement problem in quantum theory still remains an unresolved issue, with competing pro-

1. See B. Greene, *The Elegant Universe*, Jonathan Cape, 1999.

posals being made that correspond to radically different kinds of metaphysical picture, none of which is altogether free from difficulties.

Considerations of this sort mean that it is not instinctive for a scientist to adopt an attitude that leads to asking about a fundamental theoretical proposition the question, 'Is it reasonable?', as if one expected to possess beforehand knowledge of the form that rationality has to take. The physical world has proved to be too surprising for that kind of a priori confidence to be appropriate. The fact is that the universe has often turned out to be recalcitrant to our expectation. Therefore, for the scientist the natural question to ask takes the different form, 'What makes you think that might be the case?'. This is an altogether more open form of enquiry, free from prejudgement but insisting on there being evidential motivation for the belief eventually embraced. A familiar example of the need for this style of thinking is provided by the well-known story of investigations into the nature of light. Nineteenth-century physics decisively established that light possessed wavelike properties, but ideas formulated by Max Planck and Albert Einstein in the early years of the twentieth century, and finally confirmed by a brilliant experiment performed by Arthur Compton in 1923, equally clearly showed that, in certain circumstances, light manifested a quantised, or particlelike, form of behaviour. Any first-year philosophy undergraduate in 1900 could easily have 'proved' the impossibility of something behaving sometimes like a wave and sometimes like a particle. After all, a wave is spread out and flappy, while a particle is concentrated and behaves like a bullet. For a while, wave/particle duality appeared to be a necessary nonsense in physics, demanded by experience but denied by reason, until Paul Dirac's discovery of quantum field theory dissolved the paradox. In that theory, because the fundamental entity is a field, spread out through space and varying in time, it has wavelike properties; because it is quantised, its energy comes in countable packets that behave just like particles. Under the nudge of nature—the impact of the sheer stubbornness of the way things are—a rational possibility had come to light that would have been likely to have lain undisclosed to a priori thinking.

I have called the style of argument, so congenial to a scientist, that seeks to proceed from the influence of interpreted experience to the formulation of theoretical understanding, 'bottom-up thinking'. It contrasts with a foundationalist style of 'top-down thinking', which believes that one can descend from known general rational principles to the reliable consideration of the nature of particulars. I believe that it is also possible to approach theological issues in a bottom-up fashion and I devoted my Gifford Lectures to a defence of the Nicene Creed set out along these lines.[2] The exhibition of the experiential motivations for orthodox Christian belief affords the opportunity to rebut the opinion, in my experience quite commonly held among scientists, that religious belief is based solely on an uncritical submission to some kind of unchallengeable authority. I believe that many of my colleagues in science are both wistful and wary about religion. They are wistful because they can see that science on its own is too limited a form of enquiry into reality for it to be able to answer every question that it is reasonable and necessary to ask. They

2. J. C. Polkinghorne, *Science and Christian Belief/ The Faith of a Physicist*, SPCK/Fortress, 1994/1996.

would like to discover that there is a more profound and comprehensive story to be told. Religion offers such a story, but the scientists are also wary because they fear that religious belief demands intellectual suicide—sign on the dotted line and do not ask any questions. Of course, my colleagues do not want to make so irrational a commitment, but nor do those of us who are seeking to serve the God of truth. In fact, the question of truth is quite as fundamental to theology as it is to science, and in both forms of investigation it is the search for motivated belief that is the central technique in the quest for true knowledge.

In its enquiry into the nature of divine reality, theology cannot expect to escape from problems similar to those which science has had to face in the course of its exploration of physical reality. Theology's thinking will always have to be open to revision in the light of the pressure of the way things are, and this may lead to initially surprising conclusions. In a discussion of the role of coherence as a control on theological thinking, David Brown criticises what he sees as being a frequent supposition in much orthodox discussion, that 'if the grounds for belief seem good, then the logic must be right', though he concedes that if the grounds really are good this is 'indicative of something approximating the truth', even if a logical critique may then reveal that 'a shift (not necessarily a major one) in the form of the belief must be made'.[3] For the reasons already given, I am less optimistic about unaided human powers to assess coherence. In my view, the experience of doing science encourages a delicately balanced approach that, on the one hand, does not ride roughshod over the search for coherent

3. D. Brown, *The Divine Trinity*, Duckworth, 1985, p. 221.

formulation but, on the other hand, does not neglect the possibility that new forms of rationality may come to light only under the unrelenting pressure of actual experience. Clearly acts of judgement are called for in seeking to steer a course between the twin perils of a rash fideistic assertion and a timid submission to the protocols of a purely secular philosophy. As much in theology as in science, the exercise of discernment necessitates the employment of tacit skills of a kind that cannot be reduced to following a given set of rules.[4] Whatever the rigorist and polemical second-century Christian thinker Tertullian meant by his celebrated boast 'It is certain because it is impossible', he was right to the extent that the search for theological truth cannot equivocate about the special character of its properly motivating evidence, conveyed in scripture and preserved in the life of the Church, even if it leads to an understanding that seems 'foolishness to Gentiles' (1 Corinthians 1:23). But neither can theology excuse itself from the apologetic necessity to help believers to be 'ready to make your defence to anyone who demands from you an account of the hope that is in you' (1 Peter 3:15).

If the quantum world requires its own form of logic, one might anticipate that everyday habits of thought may also require some revision when one engages in the task of seeking to understand divine reality. In addition to that general consideration, there are also particular limitations to be expected in the degree of success attainable in the specific case of theology. The infinite nature of God is never going to be exhaustively contained in the finite categories of human thought. The mysterious ineffability of the divine, emphasised by what is

4. Cf. M. Polanyi, *Personal Knowledge*, Routledge and Kegan Paul, 1958.

called apophatic theology, must always be borne in mind in the course of an honest enquiry. Yet the mystery card should be the last one to be played in theological discussion, for Christians believe that God has acted to make the divine nature known in humanly accessible ways, particularly in the life, death and resurrection of Jesus Christ.

We now need to turn to the question of what might actually be the motivating evidence held to support a Christian understanding of the trinitarian nature of God.[5] The principal resources available divide between scripture and liturgy, that is to say between the accounts of the foundational events that gave rise to the Christian movement and the continuing experience of believers that is expressed in the worshipping and serving life of the Church. I shall not here essay an assessment of the authenticity of this material, absolutely necessary though that task is, for it is something that I have attempted elsewhere.[6] I do not want to repeat that exercise on this occasion, beyond the preliminary considerations already given in the preceding chapter in relation to Jesus Christ. Accordingly I shall assume that which is taken to be fundamental by all the writers of the New Testament, that Jesus died on the cross, deserted by his followers, and that on the third day he was raised from death to an unending and glorified life. I have already made it clear that I recognise that belief in the resurrection is scarcely an unproblematic matter for contemporary enquirers. Yet we have seen that it is the pivot on which the credibility of Christianity turns, for it resolves what would otherwise be the ambiguity of that apparently miserable end on Calvary—

5. See also Polkinghorne, *Christian Belief/Faith*, pp. 154-6; *Science and the Trinity*, SPCK/Yale University Press, 2004, ch. 4.
6. Polkinghorne, *Christian Belief/Faith*, chs 5-8.

was it finally defeat, or was it a surprising and wholly unanticipated kind of victory, stranger than we could have thought? Of
course, there could never be absolute certainty about so unprecedented an event in past history, but I have indicated why
I am persuaded that the Christian conviction that 'Jesus lives!'
is by far the best explanation of how it came about that a small
band of dispirited and disillusioned followers was transformed
into those who 'turned the world upside down' (Acts 17:6).
The affirmation of its belief in the resurrection of Christ, and
of trust in its risen Lord, has been the continuing testimony
of the Church down the centuries. We have seen in the last
chapter that the writers of the New Testament, as they considered their experience of the exalted Christ, found themselves
being driven to use divine-sounding language about Jesus, despite their knowing that he was a human figure of the recent
past. As a further example of this tendency we may consider
what appears to have been the earliest form of Christian con-
fession, 'Jesus is Lord' (for example, 1 Corinthians 12:3), which
not only opposes the Roman political claim that 'Caesar is
Lord' but also attributes to Jesus a title that in Jewish thinking
rightly belonged to the one true God of Israel, whose name
was too holy for human utterance, so that it had to be replaced
by the circumlocution *Adonai*, Lord. Here is a seed from which
would grow, after much theological struggle, the distinctive
Christian beliefs in the incarnation and the Trinity. Certainly,
all subsequent Christian reflection on the nature of God has
taken place in the context of the death and resurrection of
Jesus, and of the Pentecostal experience of the pouring out of
the Spirit on that first band of followers in Jerusalem.

The terrible twentieth-century events of war and genocide have meant that much contemporary theological think-

ing is deeply concerned with the problem of suffering. It has become widely recognised that it is not sufficient simply to envisage God as a compassionate spectator of the anguish of the world, looking down from the inviolate sanctuary of heaven onto the bitterness of earthly history, but there must be a more profound divine engagement with creation's passion. One of the most powerful expositors of this point of view has been Jürgen Moltmann. In his book on trinitarian theology, he writes that 'fundamental theology's access to the doctrine of the Trinity is carried out today in the context of the question about God's capacity or incapacity for suffering'.[7] The cross of Christ is seen as the focal point of divine involvement with the travail of creation.

Moltmann has been very bold in his interpretation of the trinitarian character of the passion of Christ. He writes that 'The stories of Gethesemane and Golgotha tell the history of the passion which takes place between the Father and the Son'.[8] For Moltmann, the cry of dereliction ('My God, my God, why have you forsaken me?'; Matthew 27:46; Mark 15:34) is not simply the expression of Jesus' solidarity with humankind in the harrowing experience of the apparent absence of God, but it is also an internal event within the life of the Godhead itself. The paradoxical mystery of the cross is that here 'The Son suffers in his love being forsaken by the Father as he dies. The Father suffers in his love the grief of the death of the Son'.[9] For many of us, this profound insight that the Christian God is the Crucified God is what makes belief

7. J. Moltmann, *The Trinity and the Kingdom of God*, SCM Press, 1981, pp. 4–5.

8. Ibid., p. 76.

9. J. Moltmann, *The Crucified God*, SCM Press, 1974, p. 245.

possible in this world where there is so much that is strange and bitter. Moltmann's emphasis on the centrality of the cross to an understanding of the nature of God is powerful and sustained.

> If the central foundation of our knowledge of the Trinity is the cross, on which the Father delivered up the Son for us through the Spirit, then it is impossible to conceive of any Trinity of substance in the primal ground of this event, in which cross and self-giving are not present. Even the New Testament statement 'God *is* love' is the summing up of the surrender of the Son through the Father for us. It cannot be separated from the event on Golgotha without becoming false.[10]

Two comments may be made. One is that Moltmann's thinking, though undoubtedly trinitarian in intent, can at times seem almost binitarian in tone.[11] In his account of Calvary, emphasis is laid on the relationship between the Father and the Son, with the Spirit seeming to fulfil the somewhat background role of the medium through which relationality is sustained even in the depths of forsakenness. This 'self-effacement' of the Spirit can be seen as reflecting an element of reticence often to be found in the Christian characterisation of the nature of the Holy Spirit.[12] Representations of the Trinity in Western art routinely use the images of the kingly Father, the crucified Son, and the modest presence of the Spirit portrayed as a small white dove.

The second comment is that Moltmann's understand-

---

10. Moltmann, *Trinity and Kingdom*, p. 160.

11. But see J. Moltmann, *The Spirit of Life*, SCM Press, 1992, especially pp. 71-3.

12. See J. V. Taylor, *The GoBetween God*, SCM Press, 1972; also Polkinghorne, *Christian Belief/Faith*, pp. 146-52.

ing of Calvary as a transcendental event enacted between the divine Persons provides a strong bulwark against the most prevalent of all trinitarian heresies, modalism, the idea that the Persons correspond simply to three contrasting ways of approaching the single reality of God. Christian experience testifies to the knowledge of God as the Creator of the world (one might say, the Father 'above us'), as God made known to us in human terms in Jesus Christ (the Son 'alongside us') and as God at work in our hearts and lives (the Spirit 'within us'). This threefold skein of experience certainly forms part of the motivation for trinitarian belief, but taken by itself it easily leads to a way of conceiving of the Trinity as just representing three different modes of encounter with a single divine reality, three contrasting perspectives on what is essentially a monistic deity, so that the images of the three Persons are treated simply as symbols separately appropriate to different aspects of religious experience. The 'simultaneous' event of intra-trinitarian forsakenness at Calvary cannot be thought of in this way without doing violence to its proper nature. Relational experience requires a degree of differentiation between those who participate in it. Moltmann's insight demands our taking with complete seriousness the distinctions between the Persons, so that they are seen to correspond to the divine ontology and they are not simply three contrasting epistemic modes of knowledge of the divine.

In Christian thought, the cross is never to be considered in isolation from the resurrection. Moltmann is absolutely clear about this. 'A theology of the cross without the resurrection is hell itself'.[13] Similarly, at the start of his profoundly

---

13. Moltmann, *Trinity and Kingdom*, pp. 41-2.

trinitarian account of systematic theology, Robert Jenson declares that 'to attend theologically to the Resurrection of Jesus is to attend to the triune God'.[14] On the theme of the resurrection, the Christian tradition gives a less veiled account of the role of the Spirit. Paul, in a passage which many scholars believe to be a quotation from a very early pre-Pauline formula, speaks of Jesus as 'designated Son of God in power according to the Spirit of holiness by his resurrection from the dead' (Romans 1:4, RSV). Later in Romans (8:11), Paul affirms in his own words that the Spirit is the one 'who raised Jesus from the dead'. The resurrection is not the annihilation of the reality of the cross, but it is the confirmation of the Father's faithfulness, the vindication of the saving work of the Son, and the affirmation of the life-giving power of the Spirit. The polarity of cross and resurrection, kept separated by the silent tomb of Holy Saturday, sets the rhythm by which the pain of the old creation is redeemed and transformed into the joy of the new creation, thereby expressing the pattern of the divine salvific process of rescue and ennoblement.

The continuing appropriation and verification of these trinitarian insights has taken place within the worshipping and serving life of the Church. Jenson, when he speaks about trinitarian modes of thought, says that 'the school of this logic was the church's liturgy' and he draws out the worshipful pattern set forth in Ephesians (5:18-20): 'be filled with the Spirit . . . giving thanks to the Father . . . in the name of our Lord Jesus Christ'.[15] Christian prayer is always understood to be addressed to the Father in union with Christ and in the power

14. R. Jenson, *Systematic Theology*, vol. 1, Oxford University Press, 1997, p. 13.
15. Ibid., p. 92.

of the Spirit. Central to liturgical experience is participation in the two dominical sacraments of Baptism and Eucharist.

The account of the baptism of Jesus (Mark 1:9-11, and parallels), with the affirmation by the Father of the Son in whom He is well pleased, and with the descent upon Jesus of the Spirit in the form of a dove, is strongly trinitarian in its imagery. In the distinctions it manifests between the roles of the Persons, it too is discouraging to a modalistic manner of thought. Moltmann simply states that 'Trinitarian theology is baptismal theology'.[16] While there is evidence that an early form of baptism was 'in the name of Jesus' (Acts 2:38; 10:48; 19:5), it is also clear that very soon the trinitarian formula 'in the name of the Father, and of the Son, and of the Holy Spirit', attributed in Matthew's gospel (28:19) to the command of the risen Christ, soon became the common practice that has continued down all the subsequent centuries.

The regular worship of the Church focuses on the celebration of the Eucharist.[17] The structure of the great Prayer of Thanksgiving, which lies at the heart of the liturgy, has the trinitarian shape of giving thanks (*eucharistia* in Greek) to the Father, recalling (*anamnesis*) the sacrificial death of Christ and his resurrection on the third day, and invoking (*epiclesis*) the descent of the Spirit on the worshippers and their gifts of bread and wine. The quintessential expression of Christian praise lies in the words 'Glory to the Father, and to the Son, and to the Holy Spirit'. Trinitarian theology is based on liturgically enacted and sustained belief.

So far we have been focusing on what the early Church

16. Moltmann, *Trinity and Kingdom*, p. 90.
17. See Polkinghorne, *Science and Trinity*, ch. 5.

Fathers called the divine 'economy', the particular historical events through which God has acted to make the divine
will and nature known, together with the less dramatic but
continually affirmative experiences of the community of the
faithful. These lines of evidence form the bases on which the
bottom-up thinker seeks to build a theological understanding
of the Trinity, reflecting on what is known of creation, incarnation and the pouring out of the Spirit. One of the most celebrated aphorisms of trinitarian theology is Rahner's Rule,[18]
which states that the economic Trinity (God manifested in
revelatory acts) is the immanent Trinity (God in the Godhead
itself). I understand this to be a principle of theological realism, asserting that the nature of God, including, of course, the
triune character of divinity, is truly made known through the
acts of divine revelation within history.

The *dramatis personae* of the revelatory drama have
proved to be threefold in their mutually interacting characters, and it is the triune God who is the ground and subject
of the Church's worship. Yet the coherent character of the
God so disclosed and worshipped is such that the Christian
community has always sought to avoid the error of tri-theism.
Christianity inherited from Judaism the latter's central and
essential belief in the unity of God, so that there is one divine
will and purpose at work in the world, even though Christian
understanding of the divine nature is nuanced in a trinitarian
way. The Church proclaims its belief in the Three in One and
One in Three, not a belief in three Gods.

This affirmation of Trinity in Unity clearly places severe
demands on the theologians who attempt some reasonable dis-

18. K. Rahner, *The Trinity*, Burnes & Oates, 1970, p. 22.

cussion of its meaning and coherence. It is important to see that this intellectual challenge does not arise from some sort of speculative mystical arithmetic, but from the demand of adequacy to actual Christian experience of the kind that we have been considering. Just as the physicists had to struggle with the duality of wave and particle because that was the task that nature had imposed upon them, so the theologians have had to struggle with trinitarian insight because the encounter with the one divine reality is inexorably shaped in a way that demands triadic understanding. It forces upon us thinking stranger than we could have thought. The process begins in the pages of the New Testament, as its writers are driven to acknowledge the Lordship of Christ and the work of the Spirit in their hearts, though they know also that the God of Israel is 'the Lord alone' (Deuteronomy 6:4).

Pursuing a little the issue of divine triunity leads to another significant dimension of trinitarian thinking, as theological reflection on unity and diversity leads to a scheme of thought that places emphasis on the ontological role of relationship. One might start with Augustine. His treatise *On the Trinity* was widely influential in Western thinking over many centuries, and it is still a text to be reckoned with today. Augustine sought models drawn from creaturely experience that might help in the task of trinitarian understanding. This quest was motivated not only by a general hope of gaining insight through analogy, but also by the specifically theological expectation that, since humanity is said in Genesis (1:26–27) to be made in 'the image and likeness of God', there would be found in human nature some pale reflections of the triune structure of divinity. Augustine possessed very deep powers of introspection that enabled him intuitively to anticipate some of the insights into the complexity of the human psyche that modern depth psychology also claims to be able to identify. He discerned within the one human being a number of threefold structures, variously described as knowing, willing and being, or as will, understanding and memory (a concept for Augustine much richer than mere recollection, for it involves the contemporarily active presence of the past). Augustine also appealed to another analogy, more overtly relational in character, that involved the triad of love: the lover (the Father), the beloved (the Son) and the love exchanged between them (the Spirit). (We note, once again, that a somewhat less than fully personal image is assigned to the Spirit.) Augustine had modest expectations of the human ability to articulate the triune nature of God. He said that one ought to answer the question 'three whats?' by the response 'three Persons' simply because it was better to say something than to remain totally silent. Christians have to speak of their faith, but with an appropriate degree of theological caution. The seriousness of the apophatic warning of divine ineffability has always to be taken into account by those who attempt the explicitness of kataphatic utterance.

The modern revival of interest in trinitarian modes of thought, which has taken place in recent years, has been inspired significantly by a recovered recognition of the metaphysical importance of relationality. Philosophical concerns of this interconnectional kind can also derive some support from twentieth-century science which, through the discoveries of quantum theory and of relativity theory, moved away from the Newtonian picture of separate atoms colliding in the container of empty space and in the course of the unfolding of an absolute time, to form a picture of physical process of an altogether more relational and integrated kind.[19] A very influential contemporary theological text has been John Zizioulas's *Being as Communion*. Working within the heritage of Eastern Orthodoxy, he states that 'The being of God is a relational being: without the concept of communion it would not be possible to speak of the being of God'.[20] The direction of theological thought in the Eastern Church has principally been oriented to moving from the Three to the One, while Western Christianity has tended to start with the One and then ask the question of how a Trinity of Persons might be accommodated within that Unity.

The patron saints of relational thinking are three fourth-century Church Fathers known, because of their geographical locations as bishops, as the Cappadocians: Basil the Great, his brother Gregory of Nyssa, and their friend Gregory of Nazianzus. Colin Gunton uses their own words to summarise their thinking as involving 'a new and paradoxical conception of united separation and separated unity', so that God is 'a sort of continuous and indivisible community'.[21] A quantum physicist immediately thinks of the faint physical analogy of the non-local entanglement of the EPR effect (p. 30). The concept of what one might even dare to call a 'divine society' casts light on that fundamental Christian assertion that 'God is love' (1 John 4:16). A strongly monistic picture of deity would seem to imply a static understanding of the role of divine love in the intrinsic divine nature, along the lines of the unrelentingly narcissistic self-regard of the God of Aristotle. In contrast, trinitarian thinking can understand the divine

19. Polkinghorne, *Science and Trinity*, ch. 3.
20. J. Zizioulas, *Being as Communion*, Longman and Todd, 1985, p. 17.
21. C. Gunton, *The Promise of Trinitarian Theology*, T & T Clark, 1991, p. 9.

love as dynamically active through a ceaseless process of exchange between the Persons, and thus that love is seen to be truly constitutive of the being of deity. Of course, there is also a danger that too social a view of the Trinity might lead to a lapse into the error of tri-theism. How to safeguard against this danger is a point to which we shall return shortly.

A trinitarian emphasis on relationship is also consistent with the Augustinian notion of looking for vestiges of the image of God within human nature. Human beings are more than isolated monads, individuals caught up in some Hobbesian war of each against all. They are persons whose nature is partly constituted and expressed through a network of contacts with other persons. True humanity finds its expression through the harmonious relationships that we seek to attain with each other, acted out within the setting of a justly ordered society.[22]

The Cappadocian idea of 'united separation and separated unity' is essential to the establishment of real relationality, which requires both mutual engagement within unity and also the preservation of distinctive character. As Gunton says, 'distinctiveness, far from being the denial of relation, is its ground'.[23] In the case of humanity, our embodiment is the means through which we are made available to each other, and that is why a truly human destiny beyond death must involve the restoration of bodiliness through the divine act of resurrection. However, this human embodied mode of individuality is very different from the trinitarian conception of the nature of the divine Persons. Recognition of this difference is

22. See C. Gunton, *The One, the Three and the Many*, Cambridge University Press, 1993, pp. 219-23.

23. Gunton, *Promise*, p. 72.

the key to the way in which theologians can rebut a charge of tri-theism. Jenson writes,

> Individuals who share humanity, the Cappodocians said, are differentiated from each other by characteristics adventitious to that humanity, as short stature, Athenian ancestry, or the like. Thus what makes Paul, Peter and Barnabas distinct human beings does not belong to their humanity. But God cannot be thought to receive merely adventitious characteristics. Therefore, if there *are* individuating characteristics of Father, Son and Spirit, the analogy with Peter, Paul and Barnabas does not hold and the unwanted conclusion [tri-theism] does not follow. Indeed, the characteristics that individuate "instances" of *God* must belong to singular Godhead itself.[24]

In contrast to human bodily separation, the divine Persons are differentiated by their modes of participation within that process of mutual indwelling in the exchange of love that theologians call 'perichoresis'. This means that distinctions between the Persons are not extrinsic, like the fortuitous differences that there are between creatures, but they are said to lie intrinsically in the kinds of relationship that the Persons instantiate within the perichoretic life of the Godhead. The struggle to articulate this insight has led theologians to have recourse to a welter of technical terminology, employing terms such as 'the Fount of being', 'filiation', 'spiration', and concepts such as 'begetting' and 'procession' to express these relational differences.[25] This is not a discourse into which I feel competent to enter. To the bottom-up thinker this kind of talk can, at times, seem arid and abstract, even displaying a misplaced confidence

24. Jenson, *Systematic Theology* i, pp. 105–6.
25. For a survey of classical trinitarian theology, see C. LaCugna, *God for Us*, HarperSanFrancisco, 1991.

in human ability to categorise ineffable mystery. Yet, I suppose that the formalism of quantum field theory might look similarly abstract to those who are not mathematical physicists (and even more so would it be the case for superstring theory). In both physics and theology, intellectual resources are being strained to the limit in the attempt to speak adequately about the rich and surprising character of reality. If the physicists seem to achieve their ends more successfully than the theologians, that is simply a reflection of how much easier science is than theology, for the former speaks of a universe that in many ways we transcend and have at our disposal to interrogate, while the latter struggles to speak of the absolute transcendent reality of God, to be met with in awe and obedience and not to be put to the test (Deuteronomy 6:16).

Theological discourse on the triune differentiability of Father, Son and Holy Spirit takes place within an overarching conviction that there is one true God. The perichoretic mutual indwelling of the Persons within the essential life of the Godhead—what theologians call the immanent Trinity—provides the internal basis for understanding this divine unity. In terms of external divine self-manifestation—the economic Trinity—there has been a tradition in Christian thinking, particularly influenced in the West by Augustine, and in the East in a different way by the Cappadocians' emphasis on the mutuality of the activities of the Persons, that has wanted to insist that the works of the Trinity *ad extra* are undivided. At first sight this claim might seem to be in contradiction to the appeal we have made to the triune nature of the *dramatis personae* in the revelatory drama of creation, salvation and sanctification, and the contrasting roles therein identified with the Persons. We think of the Father as creating, the Son as redeeming, the Spirit as sanctifying. Yet, although the Father as the Fount of being is understood to be the source of created existence, we are also told that the Word is the one by whom all things were made (John 1:3) and that in the beginning the Spirit hovered over the waters of chaos (Genesis 1:2). To take another example, Moltmann's apparently highly differentiated account of the cross of Christ is not only a description of an interior trinitarian transaction, but it arises also from interpreting externally manifested and distinctive phenomena (the darkness, dereliction and death experienced by the crucified Christ) that correspond to the fact that it is the Son who lives an incarnate human life. Yet, Calvary is certainly to be understood as a great act of *God*. We read in the New Testament that 'in Christ God was reconciling the world to himself' (2 Corinthians 5:19). To deal with this problem, Western theologians invented the idea of 'appropriation', that some divine attributes or activities are fittingly assigned to one of the Persons, without thereby denying the participation of the other Persons. The bottom-up thinker may wonder exactly how successful this top-down notion has proved to be. One is reminded of Bohr's invocation of complementarity to 'explain' wave/particle duality. He said, ask a wavelike question and you will get a wavelike answer, a particlelike question and you will get a particlelike answer and—fortunately for consistency's sake—you cannot ask both questions at the same time. Essentially, this was a rephrasing of the simple fact that that's the way things are. Real understanding had to await the discovery of quantum field theory. Appropriation looks like a theological suggestion of the fitting way to phrase questions about the nature and content of divine revelation.

It certainly seems reasonable to understand that the external expression of the unity of God is mediated through the consistent action of the divine energies at work in the world. Jenson says that 'since all divine action is the singular mutual work of the Father, Son and Spirit, there is only one such life and *therefore* only one subject of the predicate "God" '.[26] I believe that it was an intuitive apprehension of this fundamentally unified divine action underlying the drama of creation, salvation and sanctification that, doubtless unconsciously, sustained the New Testament writers as they wrestled with the necessity, laid upon them by their experience of the risen Christ, to assign Lordship to him in some kind of parallel to the Lordship of the God of Israel. Because God was the Father of Jesus, and because it was the Spirit of the Son who cried in their hearts 'Abba, Father' (Galatians 4:6), those early Christians knew that there must ultimately be a consistent account of the divine nature, just as the physicists knew there must ultimately be a consistent account of the nature of light. Neither of these sets of seekers after truth would perhaps quite have expressed their convictions in the form of words, 'if the ground of belief seems good, the logic must be right', but both would give primacy to well-winnowed experience, whatever difficulties there might be in formulating theoretically how it all fitted together.

We may even dare to say that it is the internal life of the blessed Trinity that is the ground of the divine liberty to act externally in creating the world. Jenson says, 'It is God's Trinity that allows him to create freely but not arbitrarily. His act of creating is grounded in the triune life that he is, but

26. Jenson, *Systematic Theology* 1, p. 214.

just so is not necessary to him'.[27] Divine intrinsic relationality is totally fulfilled in the perichoretic exchanges between the Persons, and so God's creative action is not demanded by any impulse to meet a divine need for the external supplementation of that relationship. Nevertheless the relational nature of deity is perfectly expressed *ad extra* by such creative action, the generous act of bringing into being a world which is the object of divine love. That created world is the vehicle for conveying the character of God that is made known through divine action within its history (the revelation that we speak of as the work of the economic Trinity). The story of creation, salvation and sanctification is the narrative of the acts of the triune God that provides the basis for the bottom-up thinker's approach to the mystery of the divine nature. That thinker's steps may at times be faltering, but the trinitarian path is the one to follow. Not all is exhaustively understood but, as with quantum theory, so with trinitarian theology, we know enough to be assured we are moving in the right direction. Moltmann affirms his belief in the essential primacy of narrative over philosophical formulation in the human quest for an understanding of the nature of God:

> For in the life of the immanent Trinity everything is unique. It is only because everything in God's nature is unique that in the ways and works of God it can be recognised as the origin of other things. In considering the doctrine of the immanent Trinity we can only tell, relate, but not sum up. We have to remain concrete, for history shows that it is in abstractions that the heresies are hidden.[28]

27. R. Jenson, *Systematic Theology*, vol. 2, Oxford University Press, 1999, p. 28.

28. Moltmann, *Trinity and Kingdom*, p. 190.

CHAPTER SIX

The Nature of Time: Unfolding Story

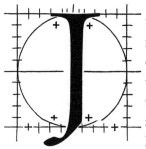 ÜRGEN Moltmann's emphasis on the role of narrative, quoted at the end of the last chapter, draws our attention to the significance of created time, within which the story of divine revelation is located and enacted. From the era of debate in ancient Greece between the followers of Parmenides and the followers of Heraclitus concerning the contrasting roles of stability and flux in the essential nature of reality, down two and a half millennia to the present day, the true nature of time has been a matter of continuing metaphysical dispute. The modern descendents of Parmenides adopt the stance that is called 'the block universe'. They believe that the true physical reality is the atemporal totality of the spacetime continuum — the whole of history, past, present and future taken together — and that human experience of the flow of time is just a trick of our limited psychological perspective as we trek along those paths through spacetime that the physicists call 'world-lines'. The

fact of the matter for these latter-day Parmenidians is that the cosmic narrative is in some sense 'already' written, and human beings are simply laboriously deciphering the text, line by line. The descendents of Heraclitus, on the other hand, believe that we live within the continuously unfolding process of a world of true becoming. The future is not 'up there', waiting for us to arrive, but we play our part in making it as we participate in the ever-developing history of the universe.

In the arguments between these two parties one encounters, once again, a debate that can be influenced by science, but which can be settled only by philosophical decision.[1] Proponents of the block universe frequently appeal to special relativity in aid of their point of view. That theory is based on assigning a fundamental role to light, understood to have the property of conveying a signal whose velocity is independent of the state of motion of the source emitting it. Of course, this postulate flies in the face of commonsense expectation. In the everyday world, a ball thrown forward from a moving train will travel faster than a ball thrown by someone standing still, as the effect of the train's motion gets added to the effect of the thrower—so why should light be different? Nevertheless, this counterintuitive universality of the speed of light has been amply confirmed by experimental testing. A consequence of this strange property is that observers' judgements of the simultaneity of distant events will depend upon those observers' states of motion, again in contradiction to commonsense expectation. We can see why this is so by thinking

1. C. J. Isham and J. C. Polkinghorne, 'The Debate over the Block Universe' in R. J. Russell, N. Murphy and C. J. Isham (eds), *Quantum Cosmology and the Laws of Nature*, Vatican Observatory, 1993, pp. 135–44; J. C. Polkinghorne, *Faith, Science and Understanding*, SPCK/Yale University Press, 2000, ch. 7.

about the following example: Consider a flash of light emitted at the midpoint of a spaceship just as the vessel is passing an observer standing on planet Earth, and ask at what times the light is reflected by two mirrors placed at opposite ends of the ship, one (B) in the bow, the other (S) in the stern. The answer you will get from an observer O_1 travelling on the spaceship is that the light hits B and S at the same time since, from that observer's point of view, it has the same distance to travel in either direction. However, from the point of view of an observer O_2 on Earth, since S is moving nearer while B is moving farther away, *given that the speed of light is the same whatever the circumstances*, the answer is that the flash will reach S before it reaches B. In other words, reflections from the two mirrors will be simultaneous for O_1, but they will occur at different times for O_2. All physicists agree that the experience of time will be different for the two observers, but there is disagreement about what metaphysical conclusion might be drawn from this.

The block theorist says that since the same pair of events could be judged either simultaneous or as occurring at different times, depending on who observes the process, it must follow that time differences are not actually significant and so equal reality must be assigned to past, present and future. The temporal theorist disagrees. Any observer's judgement about distant events is always a *retrospective* matter, since there can be no knowledge of such events until a signal is received conveying the information. It is a consequence of relativity theory that when this signal has been received, the event itself is unambiguously in the past (technically, the event then lies within the recipient's past lightcone, and the characterisation of that lightcone is independent of the observer's state of motion). In other words, judgements of simultaneity refer only to how observers organise their descriptions of the unalterable past, and therefore arguments based on such assessments can do nothing to establish the reality of the still-anticipated future.

Another argument sometimes advanced in defence of the block universe is to point to the failure of physics to incorporate into its account of nature any representation of 'now', the present moment. Since there is no preferred state of this kind identified in the physical formalism, the argument goes that the human impression of fleetingly dwelling in that present moment must be an illusion. 'So much the worse for physics', one might say in reply, 'if it proves incapable of accommodating so basic an element of the human encounter with reality'. Only someone committed to a crassly scientistic reductionism, believing that physics is all, could attempt to use such an abstract argument to discredit so basic a human experience. One might also introduce a further scientific point into the discussion. While special relativity relates perception of time to the motion of the observer, and so declines to define a universal time that might give meaning to 'now', when this particular universe is considered as a whole there turns out to be a natural 'frame of reference' (as the physicists say) that can be used to define a meaningful cosmic time. (The frame is defined by being at rest with respect to the univeral cosmic background radiation and its existence reflects the fact that, on the largest scales, our universe is effectively homogeneous.) Cosmologists are using this definition of cosmic time when they say that the universe is 13.7 billion years old. It might seem pretty far-fetched to suppose that a cosmic concept of this kind could bear any relation to terrestrial human experience, but there is another example of what appears to be some form of linkage between the universal and the local. The insight is called Mach's Principle and it draws attention to the fact that the way matter behaves in our neighbourhood correlates with the overall distribution of matter in the universe as a whole.[2]

Proponents of a developmentally unfolding account of temporality do not only appeal to the basic human experience of the flux of time, but they also point to the causally complex and fruitfully emergent picture of physical process that was discussed in chapter 2. I must confess myself to be of their party. A world characterised by sequential becoming does seem to be one that it is appropriate to consider as a world of intrinsic temporality, exhibiting an ontological contrast between the fixity of the past and the openness of the future. However, a slightly tricky philosophical point also needs to be made. It is important to recognise that issues of temporality and issues of causality are, in fact, logically distinct from each other. To equate an atemporal world with determinism is to make a category mistake, for the fact is that there is no ineluctable inference to be made from atemporality to strict determinism, or vice versa. No unique pattern of causal relationships is demanded by the events of the block universe, and what those connections might actually be is a question quite separate from that of the undifferentiated existence of past, present and future. While it would have been quite natural to think atemporally about Laplace's mechanically determin-

2. Putting it more formally, local inertial frames of reference, which refer to the dynamical properties of matter in a terrestrial environment, are found to be at rest or in uniform motion with respect to the fixed stars, i.e., the distribution of matter in the universe. (That is why the period of the Earth's rotation as measured by a Foucault pendulum is the same as the length of the sidereal day.)

istic world, in which the past and future were implicit in the present, it would not have been a forced move to do so.

Yet there do seem to be different theological implications that might be drawn from the two different pictures of the nature of time. They concern the Creator's relationship to creation. God will surely know things as they actually are, in total accordance with their nature. This seems to imply that divine knowledge of a fundamentally atemporal world would be atemporally apprehended, but a temporal world would be apprehended temporally, that is to say its events would not simply be known to be successive, but they would be known in their succession. Atemporal knowledge is precisely how classical theology thought of God's relationship to creation, believing that the whole of history is known by the divine Observer *totum simul*, all at once. All events were held to be 'simultaneously' present to the God who looks down on history from outside of time. Temporal knowledge, on the other hand, implies a true divine engagement with unfolding time. God's creative act must then be understood to have involved the gracious divine embracing of the experience of time, the acceptance of a temporal pole within divinity. This picture seems to correspond closely to how God is portrayed in the Bible, interacting with the history of Israel and accepting a radical experience of temporality in the incarnation of the Son. This insight of divine temporality, coupled, of course, with a continuing recognition of the existence also of an unchanging eternal pole within the nature of God, has received widespread acceptance in much of twentieth-century theology.[3] It does

3. Process theology lays particular emphasis on divine temporal/eternal polarity, but this insight is found also in many who are outside the process fold. Pro-

not subvert the orthodox Christian distinction between the Creator and creation, since divine temporal polarity can be understood as a form of relationship to creatures freely accepted by God as part of the process of creation, and not simply imposed upon the divine nature.[4] The concept combines naturally with an understanding of divine knowledge as having the character of current omniscience (knowing now all that is knowable now), rather than an absolute omniscience (knowing all that will ever be knowable). This restriction would be understood theologically as being kenotic, a chosen self-limitation on the part of the Creator in bringing into being an intrinsically temporal creation. It would be no defect in the divine perfection not to know the details of the future if that future is not yet in existence and available to be known.

One of the sources of philosophical perplexity about the nature of time has lain in the fleeting character of our human temporal experience. No sooner do we apprehend the present than it has receded into the past. The problem was classically explored by Augustine in the *Confessions*. He wrote, 'we cannot rightly say what time is, except by reason of its impending state of not-being'.[5] Colin Gunton suggests that the way out of the dilemma is 'that time is best understood by virtue of

---

cess thought, however, regards this polarity as a metaphysical necessity, rather than a kenotic acceptance on the Creator's part of participation in temporality. For the scientist-theologians, see J. C. Polkinghorne, *Scientists as Theologians*, SPCK, 1996, p. 41. For a discussion of the approaches of Karl Barth and Eberhard Juengel to God's relationship to time, understood in the light of Christ's suffering and death and summarised in the epigram 'God's being is in becoming', see A. E. Lewis, *Between Cross and Resurrection*, Eerdmans, 2001, pp. 188-95.

4. See J. C. Polkinghorne (ed.), *The Work of Love*, SPCK/Eerdmans, 2001, pp. 102-3.

5. Augustine, *Confessions*, XI, 14.

what takes place in it'.[6] He encourages us to approach temporality by means of the narrative of unfolding events. Someone once said that time is what stops everything happening at once, and the sequential encounters with reality that temporality enables afford a clue to how to tackle the problem of time. A religion like Christianity, that places great emphasis on historically enacted revelation, should find this approach congenial.

A wholly atemporal deity could, presumably, only fittingly be known through timeless moments of illumination. On the other hand, a God who genuinely engages with time can be known through an unfolding story of historical disclosure. That knowledge will be conveyed through revelatory events and personal encounters, rather than through the communication of timeless truths in propositional form. In a word, a theology that takes temporality seriously will have narrative at its heart. It will understand time 'by virtue of what takes place in it'. Trinitarian thinking about the economy of divine revelation, of the kind explored in the preceding chapter, provides an example.

Many early Christian thinkers looked, as we did earlier, to the narrative of the baptism of Jesus, with the distinctive roles enacted by Father, Son and Holy Spirit, as a particularly clear example of differentiation within the economy, distinguishing the actions of the Persons. The heavenly voice of the Father proclaims the beloved nature of the Son, whose solidarity with humanity is being manifested through his acceptance of John's baptism, while the Spirit descends upon him in

6. C. Gunton, *The One, the Three and the Many*, Cambridge University Press, 1993, p. 82.

the form of a dove (Mark 1:9-11 and parallels). Robert Jenson emphasises that the unity of God is expressed through the *mutuality* of the working of the Persons and not through their identity or indistinguishability. He believes that the Cappadocians got this right in the way that they thought about divine economy, but Augustine, who thought it was a matter of indifference whether the heavenly voice at the baptism was thought of as the utterance of the Father, or of the Son, or of the Spirit, got it wrong. Jenson comments that it is necessary that 'the prejudice is entirely overcome that there can be no eventful differentiation in God, *that nothing smacking of temporality can be found in him*'.[7]

Acceptance of the role of temporality in God's making the divine nature and purposes known sheds light on how to approach the use of scripture.[8] The revelation that it conveys will be the result of an evolving and cumulative record of encounter with the reality of God. The Hebrew Bible presents a picture in which clear development takes place between the time of the Patriarchs and the time of the commissioning of Moses to the task of leading Israel out of slavery in Egypt (Exodus 6:2-3, where God says, 'I appeared to Abraham, Isaac and Jacob as God Almighty, but by my name "the Lord" I did not make myself known to them'), and further development again in the course of the long ministry of prophetic encouragement and warning (Isaiah 42:9; 48:6-8). For the Christian believer, God's decisive revelatory act is the life, death and

---

7. R. Jenson, *Systematic Theology*, vol. 1, Oxford University Press, 1997, p. 110, my italics.

8. See also J. C. Polkinghorne, *Reason and Reality*, SPCK/Trinity Press International, 1991, ch. 5; *Science and the Trinity*, SPCK/Yale University Press, 2004, ch. 2.

resurrection of the incarnate Son, recorded in the New Testament but carrying a significance that is not fully expressed within its pages alone, so that the Spirit has continued the work of revelation over the centuries that have followed (cf. John 16:12–15). The community within which the Spirit works is the diachronic community of the Church, spread out in time and interconnected through its successive generations. In speaking of the hermeneutical issues that challenge contemporary interpretation of ancient scripture, Jenson says, 'our present effort to understand a handed-down text cannot be hopeless, since it is merely the appropriation of a continuing tradition within which we antecedently live'.[9] The Church's ceaseless exploration of scripture provides the context within which the power of the Bible to speak across centuries and cultural differences is actually experienced.

Such a temporal and evolutionary picture of divine disclosure demands something like an understanding of the principle of the development of doctrine. Not all is apprehended at once. The search for an adequate understanding of the significance of the foundational events of New Testament times led eventually to the trinitarian and incarnational insights of the Councils of Nicea, Constantinople and Chalcedon, and it continues beyond them in the unending task of Christian theological exploration. The time-based character of revelation makes it intelligible why there are changing theological perspectives contained within scripture itself. The acts of war and genocide that figure so largely in the annals of the deuteronomic history (Joshua to 2 Kings) do not have parallels

9. R. Jenson, *Systematic Theology*, vol. 2, Oxford University Press, 1999, p. 280.

in the pages of the New Testament (except, perhaps, for the highly charged symbolism of Revelation). The God who in the Hebrew Bible sometimes proves very dangerous to encounter (Exodus 4:24-26; 2 Samuel 6:6-8) comes to be recognised as the faithful and loving God and Father of our Lord Jesus Christ. This revisionary process continues beyond biblical times. After eighteen centuries, the Church finally came to realise the repugnance of slavery and to question whether a loving God would exact the punishment of everlasting torture for finite transgressions.

The story of the evolution of the universe, the story of the development of terrestrial life, the story of God's revelatory acts to Israel, in Jesus Christ, and to the Church, all these narratives testify to the fruitfulness of temporal process. One might think that Heraclitus has won the battle, but the ghost of Parmenides (and of Plato) is not easily laid. Many theologians have displayed a distrustfulness of time and a hankering after ultimate atemporal stability. The 'fulness of times' has been pictured, ironically enough, as a timeless fusion of all the aeons of history, and the fulfilment of human destiny has been seen as occurring in an atemporal event of illumination conveyed through the Beatific Vision. I think that this attitude is a mistake, for it denies the fact that human beings are intrinsically temporal in their character.[10] Contemplation of the divine work of an evolving creation strongly suggests that God has chosen to act through unfolding process. Yet, ultimate fulfilment cannot come simply through the continuation of the history of this world. Evolutionary optimism of that kind is an

10. J. C. Polkinghorne, *The God of Hope and the End of the World*, SPCK/Yale University Press, 2002, pp. 117-22.

illusion. Not only do all humans die to this world, but also reliable scientific predictions foresee that, after immense spans of cosmic history, the universe itself will subside into futility, either through collapse or decay. Present process, even if extrapolated far into the future, cannot fittingly be more than the first chapter of the theological narrative.

The second and final chapter of that tale has to be written in the language of eschatology, expressing hope for the life of that new creation that has already begun to grow from the seminal event of Christ's resurrection. The eschatological future is the realm within which the divine purposes will finally and completely be fulfilled in a consummation in which God is 'all in all' (1 Corinthians 15:28).[11] The crucified and risen Christ is our principal guide to thinking about this almost unimaginable destiny. Jenson says that 'the eschatological proclamation needs the narrative of Jesus to *identify* the eschaton that in fact is proclaimed'.[12]

Our life then will be embodied, but in a different kind of 'matter' from that of this world, for it will no longer be in bondage to transience and decay. (A physicist might speculate that this new 'matter' will be endowed with such strong intrinsic self-organising powers that they will overcome the tendency to disorder that in this world is expressed by the increase of entropy decreed by the second law of thermodynamics; cf p. 26.) This redeemed and glorified 'matter' is prefigured by the resurrection body into which Jesus' corpse was transformed, leaving the tomb empty, an act implying that

11. See Polkinghorne, *God of Hope; Science and Trinity*, ch. 6; also J. C. Polkinghorne and M. Welker (eds), *The End of the World and the Ends of God*, Trinity Press International, 2000; and chapter 10.

12. Jenson, *Systematic Theology* 1, p. 170.

God will eventually have a destiny for all created matter beyond the death of the universe. The life of the new creation will be a temporal life, lived within the unfolding 'time' of that world to come, whose everlasting nature is the true meaning of the fulness of times. The hurts of this world will be healed and the distortive effects of sin will be redeemed.

> Then the angel showed me the river of the water of life, bright as crystal, flowing from the throne of God and of the Lamb through the middle of the street of the city. On either side of the river is the tree of life with its twelve kinds of fruit, producing its fruit each month; and the leaves of the tree are for the healing of the nations. Nothing accursed will be found there any more. But the throne of God and of the Lamb will be in it, and his servants will worship him; they will see his face, and his name will be on their foreheads. And there will be no more night; they need no light of lamp or sun, for the Lord God will be their light, and they will reign for ever and ever. (Revelation 22:1-5)

The new creation will be the realm of realised love, for 'Eschatological description emerges as the other side of the same event of interpretation in which the gospel ethics emerge'.[13] Unending life will never become boring but it will be endlessly enriching as the redeemed are led deeper and deeper into the inexhaustible life and energies of God, made available to us as we come to know the divine nature as it is progressively unveiled. Worship will be the characteristic state of that world to come, as depicted in those great symbolic scenes of heavenly praise in Revelation (4:1-11; 5:6-14; 11:15-19; 15:2-4; 19:5-10). Jenson offers us an ecclesiological version of this hope when he writes:

13. Jenson, *Systematic Theology* 2, p. 314.

there will be a universally encompassing liturgy, with the Father as the bishop enthroned in the apse and the apostles as presbyters around him and the redeemed of all times as the congregation and the angel-driven creation as the organ and orchestra, and the tomb of all martyrs as the altar, and the Lamb visibly on the altar, and the Spirit as the Lamb's power and perception, and the music and drama and sights and aromas and touches of the liturgy as themselves the Life who is worshipped. Let us say: there will be a political community whose intimacy is such that from the vantage of this world it could only be called delirium but just so will be perfectly ordered because the delirium's dynamism will be the perichoresis of the triune life.[14]

14. Ibid., p. 340.

CHAPTER SEVEN

The Spirit and the Faiths

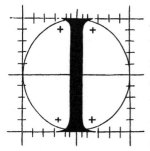AM writing from within the Christian community of faith, but one must admit that exploration of sacred reality is made problematic by the diversity of the world faith traditions. Each displays a considerable stability in its traditional heartlands. Each manifests an authenticity in the spiritual way of life that it preserves and nurtures. Today's multicultural society makes us keenly aware that this is so. People of other faiths are no longer strange people, living in faraway countries and believing very odd things. They are our neighbours, living down the street, and we can see the evident integrity of their lives. The deeply troubling intolerance and violence that some minorities within the different faith communities can display do not negate the values that are affirmed and followed by the majority.

Yet each religion also makes claims about the form of its encounter with sacred reality that appear incompatible with the testimonies of other faith communities. Is the human person of unique and abiding significance in the sight of God, as the three Abrahamic faiths (Judaism, Christianity and Islam) all affirm? Or, is humanity caught up in the revolutions of a samsaric wheel of rebirth through reincarnation, as Eastern religions believe? Or, is the human self ultimately an illusion from which to seek final release, as the Buddhist doctrine of *anatta* apparently asserts? It certainly does not seem that all these claims can simultaneously be true, as though they were simply culturally moulded expressions of the same fundamental insight.

Contradictions of this kind between the faiths are particularly disturbing to a scientist, conscious of the contrasting universality of understanding that modern science has been able to achieve. Although science started in the particular time and place of seventeenth-century Europe, it has now spread worldwide. Ask a suitably qualified person in Rome or Jerusalem, Benares or Kyoto, what matter is made of, and in all four cities you will receive the same reply, 'quarks and gluons'. Ask four people in those four cities what is the nature of ultimate reality, and their answers are likely be very divergent. Does this not show that religious belief is really just a matter of culturally shaped opinion?

I do not think so, but I acknowledge that the challenge presented by the diverse cognitive claims of the religions is one that has to be taken very seriously. Given the complexities of each of the faith traditions, and the fact that each requires for its true understanding a close degree of engagement with its life and thought, so that external inspection is insufficient, the attempt to make progress in addressing these issues of interfaith relationships is going to be an ecumenical task for the third millennium, rather than something that can be expected to be achieved by the end of the twenty-first century. The interrelationship of the faiths is a topic of very great significance, and I have tried to wrestle with it in earlier writings.[1] I only have one or two further points to make here, and the brevity of this chapter simply indicates that its main role is to serve as an acknowledgement of the importance of the issues, rather than its pretending to be able to present an adequate approach to the problems. All I can do is to acknowledge this importance and briefly sketch, from the point of view of a trinitarian Christian, on what basis I believe that the task of mutual encounter may be begun and its challenges addressed. The problems are extremely difficult and progress in understanding is likely to be slow. I believe that initial occasions of meeting will have to address serious questions, but that immediate and direct engagement with the core defining beliefs of each tradition (What about Jesus? What about the *Qur'an?*), if undertaken prematurely, would be likely to prove too threatening on all sides for progress to be made. One possible fruitful field of encounter can be afforded by the faiths sharing their insights into how they understand the discoveries of modern science to relate to their traditional theological understandings. Some progress has already been made through a project called Science and the Spiritual Quest, in which groups formed out of scientists sharing a common discipline but drawn from different faith communities met

1. J. C. Polkinghorne, *Science and Christian Belief/The Faith of a Physicist*, SPCK/Fortress, 1994/1996, ch 10; *Science and Theology*, SPCK/Fortress, 1998, ch 7. For a survey of the attitudes of scientist-theologians generally, see J. C. Polkinghorne, *Scientists as Theologians*, SPCK, 1996, ch. 5.

to explore topics of common concern. The recently founded International Society for Science and Religion, with its world-wide membership drawn from all the major faith traditions, is another promising initiative of this ecumenical kind.

In approaching these issues, I reject the notion that we Christians know everything and that all the others are just plain wrong. The spiritual authenticity displayed in the lives of my brothers and sisters in the other faiths makes this a view that it seems to me impossible to maintain. One finds oneself confronting a perplexity that insists on being taken seriously but which will not yield to being easily answered. Religious people can no more neglect the challenge of the multiplicity of the world faiths than the physicists could neglect light's duality of wave and particle.

I also reject the search for a kind of lowest-common-denominator world religion, based on what can be discovered to be shared between all the faiths. Of course, they certainly do have some beliefs and experiences in common.[2] For example, all endorse the value of compassion, and all include within the spectrum of their religious experience what is properly called mysticism, a profound encounter with sacred reality in a form variously described as being unity with the One or with the All. However, simply assembling an eclectic collection of shared themes and phenomena does not seem to me to yield an account of religion of sufficient power and coherence to be comparable to that offered by the contrasting accounts of the different historical traditions themselves. The

2. A helpful exploration of degrees of interfaith congruity is given in a series of books by K. Ward, *Religion and Revelation, Religion and Creation, Religion and Human Nature, Religion and Community*, Oxford University Press, 1994, 1996, 1998, 2000.

fact is that most religious people are Jews or Christians or whatever, rather than generic believers. This does not mean that there is no value in a kind of generalised spirituality, or in the universalised insights of a perennial philosophy, but the greater part of the vitality of the religious life as it is experienced worldwide lies in the worshipful and serving practice of the adherents of the particular faiths. One of the paradoxes of the religious scene is that the specificities that generate the disturbing clashes between the understandings of the traditions are also what give to each its living power. Accordingly, the true ecumenical dialogue of the faiths can only begin as each meets the other expressing, humbly but firmly, the foundational beliefs that are the ground of its own authenticity, even if initially this commitment may enter the discussion somewhat obliquely, rather than in an encounter of head-on collision. The necessity of witnessing to the truth as one understands it is paramount here. If you were to tell me that you believe that Isaac Newton was absolutely right in absolutely every respect in his understanding of mechanics and gravitation, I would not reply, 'Don't be silly', nor would I say, 'That's very interesting and I respect your opinion'. Instead, I would try, tactfully but firmly, to explain why I believe that Newton's profound insights need modification and amplification in the light of the further discoveries of relativity and quantum theory. Things must surely be similar in the religious search for truthful belief.

This means that I, as a Christian, have to approach those of other faiths from the perspective of my trinitarian beliefs, although I know that these are precisely what will eventually bring into focus that which will make our meeting difficult. My attitude will be undergirded by my acceptance of testimony to the revelatory actions of the three divine Persons and it would be unacceptably disingenuous to attempt in any way to disguise my conviction of the truth of the incarnation.

Appeal to the work of the Father as the Ground of being and the Source of the order and fruitfulness of the universe makes for easy contact with the two other Abrahamic faiths. All three share an understanding of the world as God's creation and this means that each engages with science's account of that world in very similar ways. After giving a public lecture on science and religion, I have quite often been approached by Jewish or Islamic members of the audience who have wanted to tell me that they too share my views on the doctrine of creation, or on an interpretation of the order of the universe understood in the terms of natural theology. Of course, I am grateful for this degree of agreement. It would be significantly helpful to extend this particular form of interaction further by being able to take more account of contributions from the Eastern religions. The claims that have been made to discern a significant correlation between Eastern thought and quantum theory[3] have largely come from Westerners. It would be very interesting to learn more about how such issues are viewed in the indigenous cultures of Hinduism and Buddhism.

Christian belief in the Son of God incarnate in Jesus of Nazareth is, of course, a critical point of cognitive clash with other faith traditions. Many Jews recognise Jesus as a remarkable teacher. Moslems know that the *Qur'an* recognises him as a prophet second only to Mohammed. Adherents of Eastern faiths have also been respectful of Jesus, with Hindus quite often being prepared to regard him as another *avatar,* one of

3. F. Capra, *The Tao of Physics,* Wildwood House, 1975. For a critique, see J. C. Polkinghorne, *One World,* SPCK/Princeton University Press, 1984, pp. 82–3.

the many forms in which they believe that the divine has been present from time to time in human life. Yet none of these faith traditions accords to Christ the unique role of the Word made flesh. I have already said that I believe that I cannot authentically enter into interfaith dialogue as a Christian if I try to cover up my conviction of the unique centrality of Jesus. That would be as misleading on my part as it would be for a Moslem to try to conceal belief in the unique status of the *Qur'an*. Our differing beliefs in these respects are why the process of ecumenical dialogue will inevitably be long and painful, and why it has to begin with matters not too intrinsically threatening. What it will eventually lead to is not foreseeable today. What I am trying to do in this short chapter is to see how I can reconcile the specificity of my Christian conviction, with its indispensable emphasis on the role of Jesus Christ as the unique link between the life of God and the life of humanity, with my firm belief that there is spiritual authenticity present in the other faiths that evokes my admiration and respect.

The distinguished Roman Catholic theologian Karl Rahner sought to address this issue in Christological terms, with his concept of 'anonymous Christians'.[4] Although Christian theology must assert that ultimately none can come to the Father except across that unique bridge between human life and the divine life that is constituted by the Word made flesh (John 14:6), it does not have to deny that more know Christ in this life than know him by name, for he is the 'true light which enlightens everyone' (John 1:6). Rahner's idea of the veiled presence of Christ in other traditions has been heavily criticised as amounting to a patronising Christian takeover bid. I

---

4. K. Rahner, *Theological Investigations*, vol. 6, Darton, Longman and Todd, 1969, pp. 390-8.

do not think that this is fair, but I do not believe that he succeeded in finding the right way to approach the problem of expressing in Christian terms the salvific presence of the sacred in the other faith traditions.

This role, I believe, is best to be understood in terms of the activity of the Spirit. The idea has a venerable pedigree, for thinkers in the early Christian centuries sought to understand the status of the 'good pagans', such as Plato, along similar lines. In the fifth century, Leo the Great said in a sermon that 'When the Holy Spirit filled the Lord's disciples on the day of Pentecost, this was not the first exercise of his role because the patriarchs, prophets, priests, and *all the holy persons of previous ages* were nourished by the same sanctifying Spirit . . . although the measure of the gifts was not the same'.[5]

We noted in chapter 5 (p. 99) that there is a reticence in Christian speaking and in Christian imagery in relation to the Person of the Spirit. This is, in fact, the expression of a deep theological insight into the nature of the Third Person of the Trinity. From within the tradition of Eastern Orthodoxy, Vladimir Lossky wrote,

> The divine Persons do not assert themselves, but one bears witness to another. It is for this reason that St John Damascene said that 'the Son is the image of the Father, and the Spirit the image of the Son'. It follows that the third Hypostasis of the Trinity is the only one not having His image in another Person. The Holy Spirit, as Person, remains unmanifested, hidden, concealing Himself in His very appearing.[6]

5. Quoted in S. T. Davis, D. Kendall and G. O'Collins (eds), *The Trinity*, Oxford University Press, 1999, p. 18; my italics.
6. V. Lossky, *The Mystical Theology of the Eastern Church*, James Clarke, 1957, p. 160.

I find this insight profoundly illuminating. The reticent veiling of the Spirit is due to the patient process by which the pneumatological image is being formed in the assembly of those transformed by the salvific reality of the sacred. This is a process that is not confined within the limits of ecclesiological definition, for 'the wind [spirit—it is the same word in Greek and Hebrew] blows where it chooses, and you hear the sound of it, but you do not know whence it comes or where it goes. So it is with everyone who is born of the Spirit' (John 3:8). The outpouring of the Spirit drew together those who were culturally and linguistically diverse in the Pentecostal reversal of Babel (Acts 2:5-12). I believe that it has been through the reticent and hidden working of the Spirit that the divine has not been left without witness at any time or in any place. Here I find the trinitarian grounding for the recognition of the presence of spiritual authenticity among the diverse faiths and the theological undergirding of the expectation of fruitfulness in the long and demanding process of interfaith dialogue that lies ahead of us.

CHAPTER EIGHT

Evil

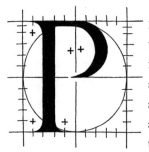 HYSICISTS are deeply impressed by the rational order and inherent fruit-fulness of the universe.[1] Many, even among those who are not adherents to any faith tradition, incline at least to a kind of cosmic religiosity of the sort that Albert Einstein expressed when he wrote of 'a feeling of awe at the scheme that is manifested in the material universe'.[2] Hence the quite frequent, almost in-stinctive, recourse to the use of 'Mind of God' language when people working in fundamental physics write books for the general public.

Biologists are different. Quite commonly they display hostility towards taking any serious account of religious ideas or language. There are at least three reasons why this might

---

1. See J. C. Polkinghorne, *Science and Creation*, SPCK, 1988, chs 1 and 2; *Belief in God in an Age of Science*, Yale University Press, 1998, ch. 1.

2. H. Dukas and B. Hoffmann (eds), *Albert Einstein: The Human Side*, Princeton University Press, 1979, p. 70.

be so. One is the unfortunate legacy of disputes over Darwin's evolutionary ideas, lingering even today in the circles of 'creationism' (so-called). We have seen already in chapter 3 that the denial of evolutionary understanding is no matter of necessity for Christian theology. In fact, quite the reverse, since respect for the truth requires Darwin's insights to be taken with appropriate seriousness. Nevertheless, the memory of some of the religious mistakes of the past lingers on in the biological community, particularly among those who take no trouble to find out what contemporary theology actually has to say.

Second, placing an extraordinary degree of overconfidence in science's unaided power to gain understanding can lead some biologists to make grossly inflated claims that their insights are capable of explaining pretty well everything. Many physicists were in this kind of grandiose mood in the generations that followed Isaac Newton's great discoveries, but the later discernment of the complex subtlety of physical process eventually led that community to a more humble recognition that mechanism is not all. Man is more than a machine. Yet biologists today, in the wake of their stunning discoveries in molecular genetics, are all too prone to a euphoric degree of unjustified triumphalism that grossly exaggerates the explanatory power of their discipline. I feel sure this is a temporary episode that will not survive a recovery of full biological interest in organisms as well as in molecules.

Yet there is also a third reason for biological reserve about religion, which is of a much more serious kind. In contrast to the austerely elegant perspective of the physicists, biologists view a scene that is much more messy and ambiguous in its character, with a mixture of fruitfulness and waste, of promise and pain. The truth-seeking explorer of reality must take this last issue with the utmost seriousness.

Of all the difficulties that hold people back from religious belief, the question of the evil and suffering in the world is surely the greatest. Narrowing the focus from nature to humanity only intensifies the issue, as the long history of war, exploitation and persecution is then brought into the perspective. How can such a world be considered to be the creation of a God who is both all-good and all-powerful? The statement of the problem is too familiar and troubling to need extensive elaboration. Not only does it give considerable pause to the enquirer after theism, but it is also one that remains a perpetual challenge and source of perplexity for those of us who are believers.

There are two different kinds of evil that need to be considered. Moral evil arises from human choices that lead to cruelty, exploitation and neglect. Natural evil arises from events outside human control, such as the incidence of disease and disaster. There is not always a clear-cut division between the two. Shoddy building practices can considerably enhance the destructive effects of earthquakes. Unjust treatment of the poor reduces their condition to an impoverished state of enhanced vulnerability to epidemics. Human lifestyle choices, such as heavy smoking, can lead to tragic early death through cancer. Yet, while the responsibility for moral evil seems to lie with human beings, ultimately the responsibility for natural evil appears to lie at the door of the Creator.

The attempt to justify the ways of God in the face of the actuality of evil is called theodicy. It is a task of considerable importance and difficulty for theologians. It is clear that the perplexities that are raised are not ones that are capable of being dispelled simply by a few paragraphs of clear-thinking prose. They are as much existential as logical and they lie very deep. Christian thought over the centuries has followed one of three basic strategies.

The first is one that the advance of science has made untenable for us today, although it was treated as very significant in the early Christian centuries. A plainly literal reading of Genesis 3:14-19 (the words of God to Adam and Eve and the serpent in the mythic story of the eating of the forbidden fruit and its aftermath) led to the idea that the Fall, understood as the original act of moral evil, also resulted in a curse upon creation that was the actual source of natural evil. Paul appears to write within this kind of understanding when he speaks of Adam as the one through whom sin came into the world 'and death came through sin' (Romans 5:12). It is obvious that our knowledge of the long history of life, with the mass extinctions that have punctuated it, does not permit us today to believe that the origin of physical death and destruction is linked directly to human disobedience to God. However, if we understand the story of the Fall to be the symbol of a turning away from God into the self that occurred with the dawning of hominid self-consciousness, so that thereby humanity became curved-in upon itself, asserting autonomy and refusing to acknowledge heteronomous dependence, we can today interpret those words in Romans in the sense of referring not to fleshly death but to what may be called 'mortality', spiritual sadness at the transience of human life.[3] Because of their self-conscious power to look ahead into the future, our ancestors had become aware that they would die. This was an emergent recognition

---

3. J. C. Polkinghorne, *Reason and Reality*, SPCK/Trinity Press International, 1991, ch. 8; *Belief in God*, pp. 88-9.

of something always present, namely the finitude of life in this world. Christian belief embraces the idea that God's purposes will find their ultimate fulfilment beyond present history in the everlasting life of the world to come, but the Fall meant that our ancestors had become alienated from the One who is the only true ground of hope for that *post mortem* destiny. Hence their feeling of the bitterness of mortality, an experience in which we also share, for we are the heirs of that fractured relationship with our Creator. This modern interpretation of the Fall and its consequences conveys an important insight into the human condition, but it does not, in itself, offer us a resolution of the problem of evil.

The second strategy of theodicy is an attempt to deny the absolute reality of evil. It is claimed that evil is no more than a kind of deprivation, the absence of the good rather than the substantial presence of the bad—rather as darkness is simply the absence of light. (There are photons, particles of light, but there are no scotons, particles of darkness.) After the terrible events of the twentieth century this seems to me to be an impossible stance to adopt. In fact, when one considers an appalling episode like the Holocaust, though one can see individual and societal factors at work (the implacably evil will of powerful leaders; a society in which an unquestioning obedience to the State had been strenuously inculcated; ordinary human cowardice that meant that people looked the other way when the cattle trucks laden with their human cargo rumbled through the village railway station on the way to Auschwitz), nevertheless there is a weight of evil involved in these dreadful events that makes me, at any rate, not quick to be dismissive of the possibility that there are also non-human powers of evil loose in the world. If that is so, it does nothing of itself to resolve the problems of theodicy, since the question of how these satanic powers originated, and why they are permitted to continue, remains deeply troubling. Whatever view one takes about the nature of spiritual evil, it seems that evil's reality is just too great to be argued away as simply the privation of the good. Yet, having acknowledged that, the light/dark comparison does serve to remind us of the existence of very much positive good in the world, so that the problem of evil has to be held in tension with the 'problem' of the existence of value and good. The world is both beautiful and ugly, inspiring and terrifying in turn.

The third strategy of theodicy is the one followed by most contemporary theologians. It seeks to make out a case that the evils that occur are the necessary cost of greater goods that could be attained in no other way and which more than redress the balance of creation in God's favour. According to this view, the dark side of creation is the unavoidable shadow that is inseparable from its goodness. In relation to moral evil, this argument is summed up in the well-known free-will defence: a world with freely choosing beings, however bad some of their choices may prove to be, is a better world than one populated only by perfectly programmed automata. This is not a claim that can be made in this post-Holocaust era without a quiver in the voice. Nevertheless, I believe that there is important truth here. We instinctively recognise that acts that seek to manipulate and restrain a person's freedom of action, even when undertaken with desirable intentions, such as various acts of restraint laid upon potential or actual offenders in order to avoid permitting the infliction of harm, are in themselves acts of imperfection, in that they diminish the humanity of those on whom they are imposed. Philosophers argue whether or not it would have been possible for God to have created beings who *freely* and *always* choose the good. There does seem to be a paradox in this notion. Yet there is also a problem here for Christian theology, since its understanding of the life of the world to come is precisely that it will have such a character of unremitting persistence in the good. I believe that the invulnerability of heaven to subversion through a second Fall arises from the fact that the unveiled presence of God, there revealed as the source of all good, will elicit a full and free acceptance of the divine will.[4] There is clearly a difference in moral status between initial imposed necessity and eventual voluntary acceptance.

But, if that is a correct view concerning the resurrection life of the new creation, why should a clear manifestation of God's goodness not be made also in the course of the life in this world, rather than waiting until the next? Putting it bluntly, why does not God make the divine will and winsome nature absolutely clear right now, that is, as soon as possible? I believe that the answer lies in the recognition that God's creative purpose is necessarily a two-step process. The first step is this present creation, existing at some epistemic distance from its Creator, whose divine presence is currently veiled from our sight. If there is truly to be an exercise of creaturely free-will, this seems to require such an initial distancing from the overwhelming presence of the divine. An initial veiling of the full revelation of God's infinite nature seems necessary if finite creatures are to be allowed a true freedom to be themselves. It is only after a free decision has been taken

4. J. C. Polkinghorne, *The God of Hope and the End of the World*, SPCK/Yale University Press, 2002, p. 134.

to renounce the illusion of human autonomy and to embrace the reality of heteronomy that the nature of God can progressively begin to be revealed with greater clarity and without forcing the individual. The encounter of the finite with the Infinite has to come about by stages.

A somewhat similar appeal to the necessity of a distance between the Creator and creation can be made in relation to the problem of natural evil. Rather than this world being a ready-made divine puppet theatre, we have seen that its character of being the home of an evolving process can be understood theologically as showing it to be a creation in which creatures are allowed 'to make themselves'. This seems indeed to be a great good, but it also has a necessary cost. As the generations succeed each other in the course of evolutionary process, death is seen to be the prerequisite of the possibility of new life. The history of the shuffling exploration of potentiality will inevitably have its ragged edges, for there will be developmental blind alleys and extinctions, as well as unfolding fertility. Another way of putting the point is to frame what I have called 'the free-process defence':[5] all of created nature is allowed to be itself according to its kind, just as human beings are allowed to be according to our kind. As a part of such a world, viruses will be able to evolve and cause new diseases; genes will mutate and cause cancer and malformation through a process that is also the source of new forms of life; tectonic plates will slip and cause earthquakes. Things will often just *happen*, as a matter of fact, rather than for an individually identifiable purpose. The question so often asked of a minister by those who are in great trouble, 'Why is this happening to

5. J. C. Polkinghorne, *Science and Providence*, SPCK, 1989, pp. 66–7.

me?', may sometimes have no answer beyond the brute fact of occurrence.

Science can offer some help to theology here in support of the necessary cost of a world allowed to make itself. We tend to think that had we been in charge of creation, frankly, we would have done it better. We would have kept all the nice things (fruitfulness and beauty) and got rid of all the nasty things (disease and disaster). However, the more science enables us to understand the nature of evolving fertility, the more we see that it is necessarily a package deal, an integrated process in which growth and decay are inextricably interwoven as novelty emerges at the edge of chaos. The ambiguous character of genetic mutation, both the engine of evolutionary fruitfulness and the source of malignancy, illustrates the point.

A theologian would say that what is involved in the occurring costliness of creation is the divine permissive will, allowing creatures to behave in accordance with their natures. Bringing the world into being was a kenotic act of self-limitation on the Creator's part, so that not all that happens does so under tight divine control. The gift of Love in allowing the genuinely other to be is necessarily a precarious gift. I believe that God wills neither the act of a murderer nor the incidence of an earthquake, but both are allowed to happen in a creation given its creaturely freedom.

There may seem to be something very bleak in such a conclusion, but I think that it represents the necessary primal reality of a world not yet fully integrated with the life of God. The free-will and free-process defences are just two sides of one coin, the cost of a world given independence through the loving gift of its Creator. The two insights are also linked by the fact that the possibility of the morally responsible exercise of free will depends upon its taking place in a world of sufficiently stable integrity that actions can have foreseeable consequences. The ethical imperative of care for others would become meaningless if God could always be relied upon for magical interventions to save people from the bad consequences of human carelessness and neglect.

In making arguments about theodicy, the Christian has to exercise great discretion in appealing to the life of the world to come. A facile invocation of future good as the means of explaining away present ills can be insensitive and unconvincing. In Dostoevsky's novel, Ivan Karamazov was right to insist that it is not acceptable simply to regard the intense suffering of a boy, painfully and unjustly put to death, as the justified price for bliss to come. That kind of transactional argument, simply stated on its own, is callous and immoral. But it is still true that the boy's fate is yet more tragic if he has no destiny beyond his terrible death. Whatever value the insights of theodicy may have, they are a kind of interim judgement on present process, and the theological account is incomplete unless it is perceived also to affirm the eschatological hope of the ultimate absolute triumph of good over evil. The first step of God's creative activity represented by this present world is indeed a precarious venture, and it needs for its final fulfilment and justification the second step of God's redeemed new creation. The cry *O felix culpa!* expresses the belief that nothing is beyond God's final power of rescue and renewal.

Part of the problem of evil is simply its scale. Some degree of danger and struggle could be seen as providing a challenging spur to growth and development, but too often suffering seems only to diminish or extinguish the humanity of those on whom it falls. There is a mystery here that will not yield simply to rational analysis. In reality, the problem of evil is too profound to be dealt with adequately by any form of moral bookkeeping, as if one were simply casting up creation's ethical profit-and-loss account. Much of the discourse of philosophers on this issue, whether of theistic or atheistic stripe, is too coolly detached to carry much conviction.[6] The precise quantification of evil is a highly problematic notion, even if one can see that there are greater and lesser ills.

Ultimately, responding to the surd of tragedy requires the insights of the poet more than the arguments of the logician. I have already indicated how important for me is the passion of Christ, understood as divine participation in the travail of creation (p. 98). Here is a point unique to Christianity, with its trinitarian and incarnational understanding of the nature of God. One might dare to say that the burden of existential anguish at the suffering of the world is not borne by creatures alone, but their Creator shares the load, thereby enabling its ultimate redemption. Christianity is a religion that often calls for the acceptance of suffering, in contrast to the Buddhist counsel to flee suffering, and it does so because it can speak of that acceptance as a participation in the sufferings of Christ (1 Peter 4:12-19). The Christian God is the crucified God, not a compassionate spectator from the outside but truly a fellow sufferer who understands creatures' pain from the inside. Only at this most profound level can theology begin truly to engage with the problem of the evil and suffering of this world.

6. See, for example, E. Stump and M. J. Murray (eds), *Philosophy of Religion: The Big Questions*, Blackwell, 1999, pp. 151-262.

CHAPTER NINE

Ethical Exploration: Genetics

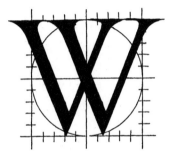RESTLING with the problems of good and evil takes two forms. One is the general, and often seemingly rather abstract, consideration that concerned us in the preceding chapter. The other is the more concrete endeavour to reach right ethical conclusions about decisions at the level of individual and societal responsibility exercised in specific situations. This chapter presents a particular case study of a set of ethical issues that arise from recent scientific and technical advances.

Science gives us knowledge, a gift that is surely always welcome as providing a better basis for decisions than ignorance. But then science's lusty offspring, technology, uses that knowledge to give us power, the ability to do things not previously thought to be possible. This is a more ambiguous gift, since not everything that can be done, should be done. Therefore humanity needs to seek the further gift of wisdom, the ability to discern and choose the good and to discern and refuse the bad. In the literature on science and religion, with some honourable exceptions,[1] less attention has been paid to ethical issues than to theological ones. I want here to explore some of the ethical perplexities that arise from the application of scientific discoveries, using the consequences of recent spectacular advances in genetics as the source of examples for this engagement with moral reality.

In order to develop the ability to make right decisions about the use of power, one might hope to gain some help from religion, for the faith traditions are reservoirs of experience and wise insight in relation to questions of ethical decision, accumulated over centuries. However, religions in their turn have to be willing to recognise that the advance of modern science has given rise to moral dilemmas that have no clear precedents in the problems of the past. For the Christian, the Bible expresses important general insights into moral principles and into God's will for creation, but its pages do not convey in any direct fashion the answers to many of the novel ethical issues of today.

Many of our ethical perplexities arise not from uncertainty about basic moral principles, but from a difficulty in understanding how these principles are actually to be applied in specific circumstances. Such problems are specially likely to arise when scientific advances take us into realms of possi-

---

1. Examples include: I. G. Barbour, *Ethics in an Age of Technology*, Harper-Collins, 1993; D. Bruce and A. Bruce (eds), *Engineering Genesis*, Earthscan, 1998; C. Deane-Drummond, *Biology and Theology Today*, SCM Press, 2001; T. Peters, *Playing God?*, Routledge, 1997. Jürgen Moltmann has also shown a deep ethical concern about ecological issues in many of his writings.

bility of a wholly novel kind. Hard thinking and frank debate will then be required to discern the right way ahead.

In a situation of this kind, careful public discussion is essential, preferably taking place before the new technology is actually on the shelf and coming into use. Genies are not easily put back into bottles. The participation of the relevant experts in the debate is indispensable, for only they can tell us what has become possible and what its further consequences might prove to be. Yet, they cannot be left to act as judges in their own cause. As a Charles Wesley hymn reminds us, every calling has its snare. The danger for the researcher is that of being carried away by the excitement of the research: 'We've done this, we've done that; come on, let's do the next thing'. But not everything that can be done, should be done. Hence the role of society in the discussion, collectively seeking the wisdom to choose the right and refuse the bad in the course of its use of the technological power that stems from advances in scientific knowledge. However, too much current ethical discourse takes the form of the clash of single-issue pressure groups, one proclaiming 'X is wonderful; let's get on with it', the other proclaiming 'X is disastrous; don't have anything to do with it'. Whatever X may be, it is very unlikely that either of these statements is true. X will be good for some purposes and bad for others. Hence the need for temperate and rational moral debate.

These general points can be illustrated by a specific consideration of problems arising from recent rapid advances in human genetics. We turn first to issues relating to the status of the very early embryo, concentrating initially on the question of the ethics of artificial cloning. Involved is the technique of cell nuclear replacement (CNR), whose use led to the birth of that most famous of all sheep, Dolly. The nucleus of a cell taken from a sheep's udder was transferred into an enucleated egg cell, which after cooling and electric stimulation was (by a process not yet fully understood) induced to become an embryonic cell that could be implanted and grown to term in the womb of a host ewe, so leading to the birth of a genetic clone of the original sheep. The issue to be considered is the possible application of CNR in the human context.

It is very widely agreed, with only a very few rogue exceptions, that *reproductive human cloning*—the strict analogue of the process that led to Dolly—is ethically unacceptable today. At the present time, the use of CNR is of very low efficacy. Two hundred and seventy-seven attempts were necessary to produce Dolly. There would also be grave safety problems to be expected in view of the many failures and malformations found in the animal experiments. In addition, there are the long-term health uncertainties suggested by the premature ageing that led to Dolly's early death. It would be extremely irresponsible to take these risks in the case of human subjects. Even if these difficulties were eventually found to be surmountable, serious ethical doubts would remain. The existence of the natural clones that occur within the same generation by the birth of identical twins is sufficient to disabuse us of the notion that copying genetic identity would in itself be a violation of personal individuality. And the personal distinctions that exist between identical twins also show that an attempt to use reproductive cloning in an effort to 'replace' someone lost through death would be to act on a fallacy.

In a quite different kind of proposal, it has been suggested that CNR might be the final resource for an infertile couple seeking a genetically related child, with the father supplying the nuclear DNA and the mother the egg, with its small amount of mitochondrial DNA. However, this proposal raises the issue of the moral propriety of transgenerationally creating a child who was essentially the identical twin of his father—a totally unprecedented and highly questionable form of family connection that might run the risk of creating seriously disturbed relationships. Of course, such a child would certainly be wanted and loved, but surely that love would be put under unusual strain if the father were tempted, as well he might be, to project his own hopes and fears about himself onto his developing clone-son. Moreover, the decision to determine another person's genome, whether in this manner or by any other form of genetic engineering, would seem to be an act of extreme prior manipulative power exercised upon them, which could be seen as a violation of their personhood through the imposition of another's will in trespass on their freedom. One can acknowledge the sense of bereavement felt by many couples unable to have children who are their genetic kin, but there are surely moral limits to what may be done to help them. Society has a role in setting these limits. In the United Kingdom, any attempt at reproductive human cloning would be a criminal offence.

On the other hand, research into so-called *therapeutic cloning* has recently been made legally permissible in the United Kingdom on a licensed case-by-case basis. The aim of this work would be to use CNR to produce a very early embryo with the same genetic make-up as a specific living person. The intended purpose would not be the illegal implantation of this embryo in a woman's womb to produce a clone of that living person, but the development of the embryo *in vitro* to the blastocyst stage (five days) and the removal of totipotent stem cells from it, so that they could be cultured and induced to yield specific tissue cells for use in the experimental treatment of serious medical conditions, such as Parkinsonism or Alzheimer's. Because of their cloned genetic nature, these cells would be immunologically compatible with the intended recipient, and so they would expected to be free from problems of rejection. In the United Kingdom, licences necessary for experiments of this kind, and for other work using early embryos, can be issued by the Human Fertilisation and Embryology Authority, provided it is satisfied that the research is for serious medical purposes unlikely to be attainable by a non-embryonic route. Only embryos of less than fourteen days' age may be used in this way. Everyone is agreed that if *adult* stem cells could be reprogrammed for tissue repair, that would be a much more desirable technique, free from moral perplexity and probably ultimately much easier to implement, but many experts think that, at the least, some embryonic experimentation will be necessary in the quest to gain understanding of cellular developmental processes before this could happen successfully.[2]

Such is the legal situation in the United Kingdom today, but there has been a significant amount of argument about whether it is ethically acceptable to experiment with the early human embryo, whether that embryo has been cloned or obtained in some other way (such as being donated from among 'spare' embryos remaining unused after an IVF procedure). Law and morality do not necessarily coincide. It seems to me

2. Another ethically untroubling source of compatible stem cells would be those derived from storing blood derived at birth from the umbilical chord, but of course this is a strategy requiring long-term planning and it offers no help to present sufferers from disease.

that the discussion does not centre on disagreement about basic moral principles, but rather on how they are to be applied. It is an agreed general principle of medical ethics that interventions on a human person are to be for the benefit of that person, and that any exception to this rule (as, for example, the surgical removal of a kidney to be donated to a compatible recipient) must be subject to safeguards that have to include the free and fully informed consent of the person involved. In Immanuel Kant's famous formulation, human persons are always ends and never simply means to an end. This principle is fundamentally unchallenged, but its application is by no means always straightforward. Coming straight to the point of our present concern, how should the principle be applied to the very early embryo?

If it is correct, as the official teaching of the Roman Catholic Church affirms, that an embryo is a full human person from the moment of its formation, then it would be agreed to have the moral status that makes its destruction in order to harvest its stem cells (in the way that therapeutic cloning or other kinds of embryonic stem cell research would require) totally ethically unacceptable. But is that judgement of embryonic moral standing correct?

The UK Human Fertilisation and Embryology Act of 1990 is based on the recommendations of a prior government-appointed body, the Warnock Committee, which concluded that, while the early embryo is certainly potentially human, it is not initially fully human, with the absolute ethical status that would confer. The very early embryo is entitled to a deep moral respect because of its potential personhood, so that it is not just a speck of protoplasm that you can do what you like with and then flush it down the sink when you have finished with it, but it is not yet a full human being. One might say that it has human life, but it has not yet attained human personhood. It is on the basis of this ethical judgement that the Act permits the licensing of certain kinds of research uses of embryos before the age of fourteen days.[3] This particular time limit was chosen because it coincides with the first onset of structure within the embryo, with the formation of the primitive streak from which the central nervous system will eventually develop. Before fourteen days the embryo is an undifferentiated mass of totipotent stem cells, capable of developing in due course into the various kinds of tissue that make up the human body, but not yet specialised in any way. At this very early stage, it is also possible for the embryo spontaneously to divide, subsequently developing into identical twins. So there are serious arguments for the Warnock position, but the absolutely central ethical issue in relation to embryonic research is whether this judgement is in fact correct, or whether the truth lies with the official Roman Catholic position.

So, when does an embryo become a human being? The arguments about embryonic stem cell research have arisen because different people answer this question in different ways, some affirming that it is indeed at the moment of conception, others that it is only at some later stage of development. Is it possible to find some rational grounds for decision that can advance the discussion beyond the simple assertion of an intu-

---

3. This tight limit on embryonic age contrasts with UK legislation on the termination of pregnancy, which allows abortion for a wide range of reasons up to twenty-four weeks, and even up to term in the case of grave handicap. Embryo research *in vitro* and termination of pregnancy *in gastro* are not precisely equivalent situations (for example, the interests of the mother play a distinctive role in decisions about the latter), but the difference between these limits is so wide as to raise serious questions of ethical consistency.

itive feeling? I think that the attempt to answer this question returns us to the fundamental question of the intrinsic nature of the human being and the meaning of the soul (p. 47). Different answers on this issue will lead to different evaluations of the status of the early embryo.

A dualist view of humanity treats the soul as a spiritual entity additional to the physical body. In the Christian context, one will then see the soul as a once-for-all spiritual endowment given by God, and in some ways it is quite natural to identify the moment of its bestowal with the moment of conception (though it must be said that no church holds requiems for embryos that die through failing to implant,[4] and we have seen that before fourteen days the embryo is not even potentially linked with just a single human person, since there is a possibility that it will divide to give identical twins). If that understanding of ensoulment at conception is correct, the embryo will be fully human from the start and its destruction to provide a source of stem cells would be as morally unthinkable as would be the removal of a living person's heart to provide an organ for transplantation into another body. Thus a dualist view of human nature accords naturally with taking a rigorist view against the ethical acceptability of embryonic experimentation and therapeutic cloning.

4. The number of embryos that fail to implant is substantial, and may be as high as 75 percent. If these are all fully human, the world to come will have a large population of persons who have never attained any degree of development in this world. Of course families understandably grieve for formed foetuses that miscarry and for stillborn children, sometimes holding memorial services for them and entertaining hope for their lives in the world to come. These practices do not demand a dualist concept of the soul, for they are consistent with a view of human nature that sees the soul as developing, combined with a trust that no attained good is ever lost in the Lord.

In contrast, a psychosomatic view of human nature, such as that which I defended in chapter 3, leads to a developmental view of the soul as something that forms and grows. Full humanity is not attained when the embryo is first brought into being, but it requires the unfolding of a process over time. The only pattern present in the very early embryo is the genome carried by each of its undifferentiated cells. On this view, the moral status of the embryo is something that enhances as the foetus develops. It is interesting to note in this regard that Aquinas (following Aristotle) held that ensoulment took place between forty and eighty days. The fact that the very early embryo has no information-bearing pattern beyond that carried by the DNA in its cells implies that the Warnock time limit of fourteen days as the period within which the instrumental use of the embryo for serious purposes would be ethically permissible can be seen to be appropriately cautious and conservatively calculated. It is in the light of this understanding that I personally have felt able, when serving on various government advisory committees, to endorse the stance now expressed in United Kingdom legislation. It is important to recognise that there is no 'slippery slope' joining therapeutic cloning to reproductive cloning. Forbidding the implantation of a genetically manipulated embryo places a barrier between the two which is well defined and clearly legally enforceable.

Before we leave the matter of embryo research, there is a further point to be made. What seemed at the start to be a very focused and specific question (Are embryonic experimentation and therapeutic cloning ethically permissible?) has turned out ineluctably to involve very fundamental issues (What is human nature? What is the soul?). We can see how essential it is that moral debate is not confined narrowly to the experts of any particular kind, but it has to be conducted in a wide forum, allowing for broadly informed participation and calling upon all the resources for understanding that are available to us.

The discussion can be briefer in moving on from cloning to other ethical issues related to human genetics. This is not because these matters are less important, for they are certainly significant, but because the thinking that is necessary has a closer connection with general considerations of established medical ethics. There are two broad areas of concern.

First, *testing and selection*. It is probable that the most extensive medical use of genetic testing and profiling will eventually be in the area of what is called pharmacogenomics. People react to drugs in different ways, both in relation to how effective the drugs will prove to be and also in the degree of severity of consequent side effects. At present, physicians simply have to try a drug and see what happens, moving on to an alternative if that proves necessary because of difficulties with the first choice. It is likely that this variety of response is substantially genetically influenced and that in the future it will prove possible to use genetic testing to determine the drug that is tailor-made to produce the best outcome for the individual patient. This very important potential increase in therapeutic efficiency does not seem to pose any ethical problems in itself though, as with all advances in medical sophistication, there are challenging moral issues about how these advances may be made available to patients worldwide on a basis of need rather than wealth. Nor do there seem to be novel ethical problems in relation to the use of genetic testing as a diagnostic resource in the treatment of actual illness. Where problems do lie is in the genetic testing of the currently-well subject.

We inherit half of our genetic make-up from each of our parents. Tests have been developed that can identify the presence of mutated genes that can have a wide variety of possible types of implication for the future development of serious disease. A dominant single-gene mutation, which only needs to be inherited from one parent for it to be expressed, such as that responsible for Huntington's disease (a late-onset condition that typically leads to severe mental degeneration and death in the early forties), gives an essentially certain diagnosis of future illness and implies a one-in-two chance that offspring will also face the same fate. In contrast, a recessive mutation, such as that responsible for cystic fibrosis, has to be inherited from both parents for the disease to manifest itself. Those who have only inherited the mutation from one parent are 'carriers', who are perfectly healthy in themselves but who have a one-in-four chance of offspring with the disease if they mate with another carrier.

Some diseases, such as breast cancer and colon cancer, generally have a complex aetiology, but there is a subset of cases in which a particular genetic mutation strongly disposes those carrying it to the development of the condition. Other diseases may have only somewhat enhanced susceptibilities due to genetic mutations, while for many others the genetic component, if present at all, is probably associated with complex patterns of interaction between many genes.

Tests have been, and are being, developed for monogenetic diseases and susceptibilities. The development of so-called DNA chips will enable the rapid use of batteries of tests of this kind. Among the ethical issues that arise are:

(1) In addition to the normal considerations of medical confidentiality, genetic testing raises special issues because of possible implications for collaterals who share in part the same genetic inheritance. If someone has been diagnosed with a severe genetic disorder carrying implications for genetic kin but declines to allow the information to be passed on to them, can a genetic counsellor properly break individual confidentiality and inform siblings of the possible risk? Some difficult decisions may have to be made in individual cases, particularly if the condition is treatable, so that there are substantial benefits possible for those made aware of their condition.

(2) The genetic tests we are considering are being offered to persons at risk who currently are healthy but for whom a positive result would carry the implication of future serious illness. Particularly in the case of untreatable conditions, there is surely a 'right not to know' that people may wish to exercise after appropriate counselling prior to a decision about whether to be tested. Before a test for Huntington's became available, those known to be at risk because of family history were asked if they would want to be tested if this became possible. About 80 percent said that they would, but when a test was actually developed, the take-up was less than 20 percent. The right not to know has also to be taken into account in decisions about passing on information about genetic diagnoses to collaterals.

(3) Prenatal diagnosis (PND) is the testing of a foetus in the womb to see if it has a serious genetic disorder. If the test is positive, the parents are likely to be offered the choice of terminating the pregnancy. Facing such a decision is not a matter free from moral perplexity, not least because of disagreements about the ethically appropriate degree of severity

that should be required in the genetic prognosis offered. In the case of a disorder that would lead to the inevitable and painful early death of a baby, termination might be considered to be the more merciful course to take, but if this option is chosen in the case of a late-onset disease, such as Huntington's, does this not seem to imply that forty years of life are not worth having?

(4) Preimplantation genetic diagnosis (PGD) involves testing an embryo, formed *in vitro*, for a serious genetic disorder. If the test is positive, that embryo will not be implanted, though others formed at the same time and free from the mutation may be. Many see this selection as ethically a less-drastic decision than the abortion of a foetus developing in the womb, though this obviously depends upon one's assessment of the ethical status of the very early embryo. In law, there is no obligation on a woman to accept implantation of any embryo, though if the early embryo is a fully human person there would seem to be an moral obligation to seek its flourishing. Some therefore conclude that morality implies that embryos should not be formed without the intention to implant. Others disagree and, in fact, in many IVF treatments there will be embryos 'left over' that will not be implanted, in order to avoid multiple pregnancies. If that is the case, why not choose 'the best' (and perhaps make the others available for embryonic research)? Yet there is a danger that the use of that optimal phrase begins to imply an unacceptable degree of commodification of children. This point leads us on to a consideration of the second broad area of ethical concern in modern medical genetic practice:

Genetic manipulation and transfer. Modern techniques make it possible to engineer the genome in a variety of ways.

Should such techniques be used in the human case? One may distinguish two different possibilities:

(1) Somatic use. In this case, the appropriate cells in the body of a specific patient would be manipulated in order to remedy a defect that had resulted in a particular disease, such as cystic fibrosis. This intervention would affect only the specific individual treated. No new issue of principle seems to be raised here beyond those that apply to medical therapeutics generally.

(2) Germline use. Here the technique would be applied either to gametes (egg or sperm) or to an early embryo. In contrast to somatic use, this kind of manipulation would have effects that could propagate to future generations. Because of grave uncertainties about what these long-term effects would prove to be, and because of their irreversible character, there is currently a generally respected moratorium on human germline manipulation. Once again, however, one must ask the question what the ethical situation would be if these uncertainties as to safety were to be resolved satisfactorily. If we could eliminate by genetic engineering the propagation of Huntington's, should we not do so? That could be seen as remedying a defect by restoration to the norm. Would it be all that different from the largely successful campaign to eliminate smallpox through a vaccination programme?

But what about attempts at enhancement beyond the norm? Discussion of designer babies with desirable characteristics (athletes or intellectuals), or the self-improvement of the human race, is science-fiction talk today, but what if this became a feasibility, as forms of gender selection already are? Surely there are moral limits that must be placed on parental choice if its exercise is not to be in danger of commodifying children. In the prospect of genetically engineering progeny, one faces that danger in an extreme form. The human genome, in a sense the carrier of life between the generations, is an entity of such value that its manipulation is a matter of extreme ethical sensitivity. If it is right to suggest that the genome is a small component in the constitution of the soul (p. 48), it must surely be treated with sacred respect. We need to consider carefully whether it would not be a step too far for human beings to take it upon themselves to interfere with it. Here is a case where the hackneyed phrase 'playing God' may really be relevant as a moral warning.

It is not easy to decide where and how to draw the moral line on this last issue. It certainly cannot be done by a simple endorsement of the 'natural' and a questioning of the 'unnatural'. Much of routine medical practice is unnatural in a plain sense. Recall the reservations initially expressed about heart transplants, now readily accepted as a therapeutic resource, though as radically unnatural as any genetic transfer. In any case, human beings are themselves a part of created nature and our actions are part of its process.

The problems discussed so far relate principally to individual decisions and their consequences. There are also important social issues to be considered, particularly relating to the potential use of genetic testing in insurance and employment. Insurance is based on the principle of the mutual sharing of risk by those who choose to participate. Its fair operation requires the disclosure of relevant information to all interested parties. It has been customary for insurance companies to ask prospective clients for access to medical information, including the rough and ready genetic information implied by family history, as a means for ensuring that individuals do not try to profit by concealing known grave and exceptional risks to their life expectancy. This strategy seeks to protect not only the insurance company itself but also its other clients, whose premiums are the ultimate source of the moneys paid out. The advent of genetic testing, with its potentially much more extensive powers of prediction, clearly complicates the issue of what medical information should be required to be disclosed in the insurance market. To take a specific example, it is widely agreed that it would not be appropriate to require someone whose family history indicated a risk of Huntington's actually to take the test. Yet, if they had already taken it, should they not be asked to disclose the result? If the result had been negative, that would clearly be a fact to their advantage. If the result had been positive, would it not be as reasonable for this fact to be disclosed as it is for a serious heart defect to be disclosed, even if it implies enhanced premiums? Yet too great an insistence on the extensive use of genetic test results might create a sizeable genetic underclass who proved to be uninsurable, generating a particular social problem in countries like the United Kingdom in which there is a close connection between insurance and the ability to get a mortgage for house purchase. For its effective operation, mutuality depends upon a degree of shared ignorance of actual risk. If this ignorance is significantly reduced by genetic advances, it might eventually be considered necessary, at least within certain financial limits, to move from the principle of mutuality within a self-chosen group of participants to a principle of universal solidarity in shared risk that would embrace a total population. The commercial difficulties relating to implementing the latter might then require there to be some degree of intervention by the State on behalf of all its citizens.

In relation to employment, questions of mutuality and solidarity also arise. On the one hand, genetic testing may be appropriate to protect employees or the general public from hazards that might arise from genetic vulnerability, such as workers' adverse reactions to specific chemical substances, or the risk of the collapse of an airline pilot while flying a plane. On the other hand, employers must surely be expected, for the sake of the common good, to be willing to accept among their workers persons with a range of health status. It would be unacceptably discriminatory to endeavour to employ only the genetically super-fit.

A further social issue of considerable complexity, which has given rise to much discussion and disagreement, is the question of the patenting of genetic discoveries. The idea behind patenting is to strike a balance between putting useful knowledge into the public domain and affording its discoverers reasonable recompense for their labour of discovery. A classic principle is that natural knowledge itself is not patentable, but only the fruits of human invention. If a gene is isolated and then copied by an artificial process, does that turn it into an invention? The predominant European attitude has been to answer, No. In North America, however, the stance taken has sometimes been different. It is significant that those involved in the Human Genome Project itself have been strongly opposed to the patenting of gene sequences, though it is recognised that specific therapeutic processes based on the exploitation of the information given by these sequences fall within the scope of proper patentability. Life in general, and human life in particular, is perceived by many to have an intrinsic value (theologically, one would say arising from its status within God's creation) that makes the patenting of its elements highly questionable.

Our discussion of human genetics has identified a number of challenging moral issues. We conclude where we began, with an emphasis on the need for the widest and most measured public moral discussion of the perplexing issues raised by advances in science. These ethical issues, however, are not confined to the use of genetic knowledge in relation to the human individual, but they relate also to the whole created order.

In connection with these wider concerns, perhaps the most significant contemporary issue raised by the new genetics is in relation to the development of *genetically modified organisms*. Ever since agriculture began, there has been genetic manipulation by selective breeding and no farm animals or crops grown today are unmodified by human intervention. Yet modern science enables gene transfers to occur, such as between a fish and a plant, that could never come about naturally. It is important to recognise that the genetic modifications currently envisaged only relate to very small amounts of genetic material of a specific kind, often a single gene, so that the process is focused and controlled, in contrast to the largely uncontrollable results of conventional cross-breeding. At the present time almost all work of this kind is concerned with plants, animal projects being largely confined to the production of therapeutically valuable proteins for medical use.

The interest in developing GM crops, though partly arising from matters of consumer taste and marketing convenience, also stems from their possible relevance to world food supplies. Present agricultural methods would provide enough food for the whole of the current world population, were that food justly distributed. However, extrapolations into the near future indicate that by about 2020 there will be a serious world food shortage if population trends continue as expected and there is no further increase in agricultural efficiency. GM crops (with their potentiality to improve resistance to pests, facilitate new forms of weed control, enhance nutritional value, and extend the kinds of land that are capable of cultivation) are one possible component in a strategy to bring about further agricultural improvement (perhaps leading to a 'second revolution', comparable to the well-known 'green revolution'). Of course GM developments must be subject to stringent standards of food safety (which I believe to be a well-understood and carefully executed procedure)[5] and environmental protection (a much more complicated task that certainly requires the use of suitably monitored test plantings for its proper evaluation). The fair exploitation of these possibilities for the common good will also require globally responsible development in which multinational companies must be respectful of the rights and needs of small-scale subsistence farmers as well as those of large commercial agricultural enterprises. It is in relation to this last point that one might hope that religious bodies might be particularly vigilant and active. As is often the case, ethical indicators can point in conflicting directions. The incorporation of 'terminator' genes that make plants sterile would be helpful in coping with

---

5. Of course, not all agree. Yet I believe that unfounded fears may produce the danger that 'genetic modification' irrationally comes to be regarded as a term of intrinsically sinister import, in the way that has become the case for 'radiation'. Popular misconception about the latter led to consumer rejection of irradiated food, despite its greater bacteriological safety.

possible ecological problems of gene spread, but harmful to the interests of subsistence farmers who routinely keep seed garnered in one year in order to use it for sowing in the following year.

Another set of ethical and religious issues in relation to genetic modification arises from the incorporation of alien genes into familiar organisms. Would the presence of an animal gene make vegetable food unacceptable to vegetarians? Would the presence of a pig gene make the food unacceptable to Jews or Moslems? Would the presence of a human gene in food involve the eaters in cannibalism? In seeking to answer these questions two facts need to be borne in mind. One is the very limited nature of the gene transfers involved. We are not considering the creation of chimaeras. (Once, in the course of a discussion about the hypothetical case of a pig gene put into a sheep, a Rabbi said to me, 'If it looks like a sheep, it is a sheep'. He was displaying the robust Jewish common sense that comes from two millennia of arguing about detailed questions of *halakah*, right conduct.) The other point is that genes operate effectively and carry active information only in the context of the living cell. Outside that context they are just rather complex chemicals. This persuades some of us that a gene taken from a human genome would, for example, simply be 'a gene of human origin', and if transferred to another organism, it would not carry humanity with it. If that insight is accepted, it provides reassuring answers to the questions posed above. Not all will concur, however, and to allow free and informed choice it is widely agreed that labelling of the presence of GM food is ethically desirable. Some practical problems arise in carrying this out in relation to prepared food containing many ingredients. At some level it seems reasonable to operate a *de minimis* principle, not insisting of the over-scrupulous labelling of mere traces of GM ingredients.

All advances in knowledge offer possibilities for beneficial use and for harmful exploitation. In the search for the right decisions, we need rational discussion, moral sensitivity, temperate debate, and the participation of as wide a body of people of ethical responsibility and good will as possible. The latter will certainly not come solely from the communities of religious believers, but those communities have a vital part to play in assisting the search for the common good.

CHAPTER TEN

Imaginative Postscript: Some Naive Speculations

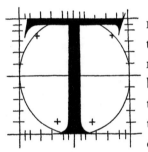

HE preceding chapters have sought to present a sober and serious exploration of reality, seeking motivated beliefs assessed in the light of trinitarian insight. Yet the investigation of theological truth can take a number of different forms. One of them is a speculative exercise in which imagination is used to pursue a theme beyond the point of firmly based knowledge. There is obviously some danger of succumbing to phantasy in the course of this endeavour, and it would clearly be an error to attempt to claim certainty where it is not to be had. Yet this kind of project, if carefully controlled and modestly undertaken, can be a useful exercise in the exploration of rational possibility. In this respect, the approach bears some cousinly relationship to the use of 'thought experiments' in science. These are mental exercises that seek to gain insight into physical principles by considering their import for idealised, and often highly simplified, situations. There was a justly celebrated sequence of encounters between Albert Einstein and Niels Bohr which was based on the discussion of such *gedanken* experiments. Einstein's purpose was to show that there was something incomplete about quantum theory and he endeavoured to demonstrate this by a series of in-principle measurements that seemed to get round the limitations of Heisenberg's uncertainty principle. Yet in each case Bohr, sometimes after a sleepless night, was able to show through careful analysis that a thorough-going application of quantum principles actually led to the confirmation of the uncertainty principle. He illustrated his replies with charmingly naive drawings of the 'apparatus' involved.[1]

In the case of theology, it is not altogether surprising that a good deal of this kind of speculative thinking has been concerned with eschatological matters. While a sufficient — and perhaps, the best — answer to many queries about the life to come may well be 'Wait and see', I do not think that the resources of imaginative thinking should be neglected altogether. Their use represents a way of investigating to some extent the coherence and credibility of eschatological hope. Some of my scientist-theologian colleagues have looked rather askance at this kind of activity, regarding an apophatic silence as being the more appropriate attitude to adopt. I certainly want to be modest and cautious in my speculative thinking. However, I believe that there is something of a loss of nerve in parts of the Christian community about affirming the hope of a destiny beyond death and that this stems partly from an inability to make some sort of sense of what that prospect might actually mean. The firm theological pillars on which Christian

1. P. A. Schipp (ed.), *Albert Einstein: Philosopher-Scientist*, Tudor, 1951, pp. 199–241.

eschatological hope rests are the faithfulness of God (Mark 12:26-27) and the resurrection of Christ. In addition, I believe that it is also important, however tentatively it has to be done, to seek to fill out somewhat the detail of the expectation that death is not the final end. In my writings on eschatology[2] I have tried to be pretty sober in exploring speculative possibility. In this concluding chapter I want to give myself a bit more rope and to speak with greater frankness about some less restrained thoughts that are in my mind as I wrestle with eschatological issues—while at the same time drawing the reader's attention to the cautionary chapter heading under which all this is written.

The discussion can be organised in the form of addressing two big questions, the first of which is: *How are the old creation and the new creation related to each other?* We have spoken already of God's creative purposes as being a two-step process (p. 142), and the issue now is how one should think about the transition from the initial stage to the final stage. Because theologically the new creation is understood already to have begun to grow from the seminal event of Christ's resurrection, the two worlds certainly cannot be conceived as being simply sequential, the second following on the total demise of the first. Instead, in some way they must exist 'side by side'. The mathematically minded will envisage this in terms of two subspaces located within the multidimensional vector space of 'total created reality'. A possibility of this kind is certainly not over the imaginative horizon of physicists. In the current version of superstring theory (a conceptual flight of fancy in its

---

2. J. C. Polkinghorne, *The God of Hope and the End of the World*, SPCK/Yale University Press, 2002; *Science and the Trinity*, SPCK/Yale University Press, 2004, ch. 6.

way as breathtaking as any idea in eschatology), it is assumed that our universe is located on a 'brane' (a multidimensional membrane) which may be only infinitesimally, but decisively, separated from other universes on other branes, which are also components in a hypothetical multiverse of such worlds 'side by side'.

The distinctive character of the old and new creations implies that our time and the 'time' of the resurrection life are, in general, associated with different dimensions of overall reality. It is conceivable, therefore, that though we shall all die at different times in this world, we shall all be re-embodied together at the same 'time' in the world to come. That would indeed be the Great Day of final resurrection. It is also conceivable that the two creations sometimes draw very near to each other, in analogy to the way in which physicists speculate about how two branes might draw close together with some form of consequent influence flowing between them. In the theological case, this might be a way to think about sacramental experience and the close presence of the risen Lord. The two creations might sometimes actually intersect, their two times briefly coinciding. I personally think in this way about the resurrection appearances of the risen Christ, thereby understanding how it could be that Jesus suddenly appeared and disappeared as the different dimensions temporarily enmeshed and then separated.

What the relationship was between the old and new creations during these periods of intersection can be explored by asking what many may think a truly naive question, Did the risen Christ breathe? If he did, there would have to have been an exchange between the matter of this world and the 'matter' of the world to come that made up Jesus' glorified resurrection body. Of course, two of the gospels in their appearance stories portray an even more drastic form of interchange, through eating (Luke 24:41-43; and perhaps, John 21:13-15a) and touching (Luke 24:39; John 20:27). Many modern theologians dismiss these details as embarrassing legendary accretions. I am not so sure. One might have expected that the earlier accounts would have been the more material in tone, with the later ones becoming more spiritual, but here the reverse seems to be the case. One detects in the contemporary assessments that are dismissive of these stories a tendency to exalt the spiritual at the expense of failing to acknowledge the abiding reality and value of the material. It seems important to hold to a middle way between a crass physicalism and a crypto-Manichean feeling that matter is only of transient significance. A similar inclination is at work when suspicions are voiced about the idea of bodily resurrection or the emptiness of the tomb. I understand the latter as arising from the transmutation of the matter of Jesus' corpse into the 'matter' of his risen and glorified body, with the important theological implication that in Christ there is a destiny beyond futility not only for humanity, but also for the whole material creation (see Romans 8:21; Colossians 1:20). Certainly, if the risen Christ breathed, there must have been some frontier of exchange at which matter/'matter' transformation took place.

Understanding that the tomb was empty because Jesus' body in the resurrection life of the new creation is the eschatological transform of his dead body in the old creation implies that the new creation is not something brought into being by God from scratch, so to speak, but it arises *ex vetere*, out of the redemption of the old.

Let me then ask another naive question. How will that

transformation come about for the great bulk of matter still remaining in the old creation, untouched so far by the resurrection? My guess is as follows. I think that God will continue to hold this present world in being while its processes are still capable of fruitful development. That fertile history will not be prematurely ended, but it cannot go on for ever because of eventual cosmic futility. If that futility comes about through the universe collapsing back into the melting pot of the Big Crunch, then that might be the epoch at which its Creator will bring its history to an end. If futility comes about through an unendingly long drawn-out dying whimper, as continuing expansion reduces the universe to extreme dilution and extreme cold (the currently favoured expectation among cosmologists), then I do not think that God will continue present history beyond the point where all possibility of significant happening has ceased. At some such moment, the matter of the dying cosmos will be changed into the 'matter' of the new creation, just as Jesus' dead body, in an unimaginable process about which the gospels are silent, was transformed into his risen and glorious body. That cosmic Day of Resurrection will be also the event in which the soul-patterns of all human beings, which have been held in the divine memory (p. 49), will be reconstituted as embodied beings living in the new creation. Human destiny beyond death and cosmic destiny beyond death lie together.

The second big question is: *What will Jesus be like when we meet him in the world to come?* Some serious theological issues lie behind this apparently naive enquiry. Orthodox Christian thinking affirms, alongside the divinity of the Son, the continuing humanity of the risen Christ. On this view, the incarnation is not a temporary episode, but an event of abiding significance. This conviction is expressed in the symbolism of the Ascension, with its picture of the human Jesus received into the cloud (not a meteorological phenomenon, but a scriptural symbol of the presence of God) so that he may take his seat in the place of authority and power at the right hand of the Father. If one believes, as I do, that the incarnation of the Son is the bridge between the life of God and the life of creatures, then our encounter with the saving reality of Christ will take place from the side of his and our humanity. We must, therefore, expect to encounter Jesus as a human figure, the pioneer of our salvation (Hebrews 2:10). The problems that this poses are not the truly naive ones of how such a vast human throng could participate in this meeting (the infinite resources of the divine must be able to cope with large numbers, however big, so that whether it is billions or trillions cannot really matter) or what language will be spoken (Acts 2:5-12 hints at the answer). The real difficulty, which may also be the clue to its solution, arises from the fact that the New Testament, though it obviously speaks of Jesus as a known human individual, also speaks of him in terms of the corporate Christ.[3] This is vividly conveyed in the Pauline image of Christians as forming part of the body of Christ (Romans 12:5; 1 Corinthians 12:27) and in Paul's frequent references to believers as living 'in Christ'. In a similar vein there is also the Johannine picture of the vine and its branches (John 15:4-7) and the fourth gospel's statement of the mutual indwelling of Christ and the believer (John 17:20-24). Even Jesus' references to himself as the Son of Man may carry corporate overtones if his

3. See C. F. D. Moule, *The Origin of Christology*, Cambridge University Press, 1977, ch. 2.

use of the phrase derives, as I believe it does, from the figure of Daniel 7, who is given a corporate identity when he is related to 'the holy ones of the Most High' (Daniel 7:13-14 and 18). Jenson frequently speaks of the *totus Christus,* Christ and the Church united.[4] The best sense that I can make of this is that our encounter with the Lord is not going to be sequential and individual, a kind of queueing up to shake hands and exchange a few words as if at some sort of heavenly Investiture Ceremony, but it will be communal and continuing. The familiar cliché of the preacher, that Christian lives are like the spokes of a wheel, the closer we get to the centre the closer we get to each other, has truth in it. Our communal encounter with Christ will be characterised both by our participation as differentiated and embodied individuals (for it will not be a stereotypical encounter, essentially the same for all) and by unity within mutual exchange (in the community of the redeemed).[5] The latter will have an almost perichoretic character of 'united separation and separated unity'. 'As you, Father, are in me and I am in you, may they also be in us . . . I in them and you in me' (John 17:21 and 23).

One final naive question remains. Suppose that self-conscious, God-conscious beings live elsewhere in the universe, as part of God's fruitful creation. Science today does not know how properly to assess the probability of there being 'little green men' out there, but theology has been wrestling on and off with what this might mean if it indeed proved to

---

4. R. Jenson, *Systematic Theology,* vol. 2, Oxford University Press, 1999, pp. 271, 289, 298-9, 333, 341.

5. Cf. Michael Welker's emphasis on the differing forms of encounter with the risen Christ in J. C. Polkinghorne and M. Welker, *Faith in the Living God,* SPCK/Fortress, 2002, ch. 4.

be the case, ever since Galileo made it plain that some of the planets are similar to the Earth. People immediately wondered whether they might not also be inhabited and to ask questions about what this might imply for the divine plan of salvation. Did Christ die for the Martians and the Venusians? Two different kinds of theological response were made. One asserted that the Son, by becoming a creature somewhere, had thereby redeemed all creatures everywhere. There is some logic to this point of view, but it seems rather hard on those other beings for them to be left ignorant of the fact of their terrestrially wrought salvation. Accordingly, there was a second kind of response which essentially said that if there were little green men in need of salvation, the Word would have taken little green flesh, just as he took our flesh for our salvation. This would accord with a particularist understanding of the famous insight of Gregory of Nazianus, that what is not assumed is not healed. It does not seem incoherent to suppose that there could be multiple incarnations in the natures of different kinds of rational beings, even if it would seem incoherent to Christian thinking to suppose that the Word lived a human life more than once. Needless to say the choice between these two responses has not been settled to universal satisfaction, but I adhere to the second point of view.

If that is right, then not only has humanity been taken into the divine life, but little green nature also. So will Christ appear as a human, or a little green man, or what? I suppose that the answer has to be 'neither and both'. Within the comprehensive community of the *totus Christus*, we shall encounter our Redeemer in the ways in which he chooses to make himself known to us. For humans it will doubtless be in the mode of authentic humanity; for others it may be in the mode of authentic greenishness. For all, it will be in the fulness of his salvific bringing together of the divine and the created. Perhaps we may also hope to encounter each other in a corresponding openness and authenticity, so that humans and little green men will come to embrace and augment each other in the endless exploration of reality which is our ultimate destiny together.

Karl Barth, arguably the greatest theologian of the twentieth century, once said that the angels would laugh when they read his theology. If they deign to read these pages, they will no doubt be in hysterics. Paul knew that sometimes one has to risk seeming a fool in order to be able to struggle to articulate something (2 Corinthians 11:21; 12:11), and that is the risk that I have taken in this short chapter.

Index

Nietzsche, F., 1, 59
non-integrable functions, 25

openness, 33–35

Parmenides, 113–14, 123
patenting, 164–65
Paul, St, 63, 83–86, 88–89, 101, 139
perichoresis, 108–109
phase space, 23
Planck, M., 92
Planck's constant (ℏ), 13, 16, 23
Plato, 123, 134
Podolsky, B., 30
Poincaré, H., 20–21, 25
postmodernism, 2
Prigogine, I., 25
psychosomatic unity, 46–47, 156

quantum theory, 12–20, 22–24,
    30–31, 43, 90–93, 106

Rahner, K., 103, 133
rationality, 92–95, 136
realism, 1–6, 34, 103
reductionism, 8–9
relationality, 104–107
Rolston, H., 49
Rosen, N., 30

Sanders, E. P., 64–66, 75, 77, 80,
    83
satisfaction, 56–57
Schroedinger, E., 14, 17

Science and the Spiritual Quest,
    129
scripture, 121–22
Shannon, C., 32
sin, 44–45
sociobiology, 53–54
Son of Man, 78–79
special relativity, 114–16
strange attractor, 27
string theory, 20, 91, 171–72
superposition principle, 18
supervenience, 37

Temple, F., 39
temporality, 39, 123–24
Tertullian, 95
theodicy, 138–45
theology, 94–96
time, 113–22
top-down causality, 27, 33
trinitarian theology, 96–112, 120–
    21
tri-theism, 103, 107–108

uncertainty principle, 12–14, 22,
    170

wave/particle duality, 92–93, 110
Wilberforce, S., 40
wisdom, 148
world faiths, 127–35
Wright, N. T., 30

Zizioulas, J., 106